Reasons Why the Theory of Evolution Isn't Just False It's Impossible

I want to apologize to the world for using strong language in this book. I am not trying to insult anyone, or bully anyone, I am only trying to convey the anger I feel against the lying devil for deceiving so many of us and the Love I feel for God who sets us free and keeps us free, by His Truth.

I was believing in the Theory of Evolution when I died as a young man, and found out the very hard way God is telling us all the Truth in the Holy Bible. So, if you have been fooled by the devil, lies in this world, don't be ashamed to admit it, we all have been fooled at one time or another in our lives. Just have the good sense to Repent of believing any lies and choose to follow the Truth, our Lord and Savior Jesus Christ, and keep following Him, all the Way Home to Heaven!

God Almighty Says So in His Word - The Holy Bible

1 Timothy 6:19-21

19 Laying up in store for themselves a good foundation against the time to come, that they may lay hold on eternal life.

20 O Timothy, keep that which is committed to thy trust, avoiding profane and vain babblings, and oppositions of science falsely so called:

21 Which some professing have erred concerning the faith. Grace be with thee. Amen.

The reason why the Apostle Paul could write that there are no actual oppositions of science to the Words of God in the Holy Bible, is that God Almighty and His Words are the reason for reality before our eyes, so nothing in creation contradicts our Eternal Creator, other than errant opinions from fallible people, lies, deceptions, and vain imaginations also known as self-delusions. So, anyone CLAIMING that science is in opposition to the Words of God in the Holy Bible is lacking accurate knowledge about science, about creation and about the scriptures, is deceived and/or self-deluded. The Words of God are harmonious in perfect synchronicity about what we

observe in His Creation because Creation isn't just Supernatural in Origin, it's Supernatural in Essence.

I'm going to lead off with the most important reason of how any of us can know Truth with all certainty; so that we are not deceived by any lies; including the redundantly disproven darwinian theory of evolution. And that reason is that God Almighty, our Eternal Creator, Lord and Savior has plainly told us the facts concerning His Creation in the provably and verifiably true contents of the Holy Bible. Yes, I know as soon as I make this statement some will roll their eyes in disbelief, but I challenge anyone who wrongly imagines that the contents of the Holy Bible are not just verifiably true and inextricably linked to reality past, present and future, but can be proven and verified as coming from the One True God, Jesus Christ; to read my books on why I say so. If you do, you will learn the most important things in the accumulated knowledge of mankind for any of us to know with absolute certainty.

(read my book on why **<u>The Holy Bible is Provably and Verifiably Divinely Inspired</u>**) - These contents specifically state that our Eternal Creator designed all living creatures to bring forth after their own kind, which means that INDENFINITELY dogs will breed with other dogs and produce only dogs, cats will breed with other cats and produce only cats, apes breed apes and produce apes, and so on so on so forth! Mankind did not "evolve" from any lesser creatures/organisms over any amount of imagined time, we are DIVINE CREATIONS MADE IN THE IMAGE OF OUR ETERNAL CREATOR (Gen 1:27) ; NOTHING ELSE; so people mate with other people and only produce other people; no matter how many imagined years by anyone might pass. (teaching that creatures evolve into completely different kinds of creatures, like the theory of evolution, is false and impossible because all living organisms are designed to bring forth only like kinds unto themselves.) And yes that is the case because God says so in the contents of the Holy Bible, anyone holding a differing opinion from those contents is wrong, lacking knowledge at best, and deceived or self-deluded at worst. In other words, NEVER believe ANYONE (scientist or otherwise) OVER GOD ALMIGHTY AND HIS WORDS, THE CONTENTS OF THE HOLY BIBLE, NEVER! even if you have a wild insane imagination like reality came into existence from nothing, because you are willfully denying the well verified facts of history, logic and reason (laws of epistemology), common definitions of words to make such an obviously false claim. Something does not come from nothing; no ifs, ands, or buts! Anyone saying so, is not in their right mind (is ignorant, deceived and self-deluded). The moment ANYONE tries to deny our Eternal Creator, the One True God, is the moment they are embracing insanity! (all such persons have no rational explanation for reality before their very eyes, and no one should listen to anything they say, which means the entire paradigm put forth by God-denying scientists today is completely false, because it comes from such state of irrational, unreasonable, ignorant, deceived and self-deluded minds! INSANE by definition and provably so!)

I feel like I wish I had a loud speaker to broadcast to the entire world: **SIT DOWN! SHUT UP! CLOSE YOUR MOUTHS AND OPEN YOUR EARS! LET THE MOST IMPORTANT INFORMATION YOU'LL EVER READ OR HEAR SINK INTO YOUR MINDS AND REMAIN EMBEDDED THERE PERMANENTLY; TOSS OUT ALL THE GARBAGE THAT HAS BEEN PUT THERE BY BELIEVING SO MANY THINGS THAT ARE NOT TRUE!**

YOU! YES, YOU! CAN KNOW OUR ETERNAL CREATOR, THE ONE TRUE GOD, WHO CREATED AND MADE THE HEAVENS AND EARTH, PERSONALLY, BEYOND ALL DOUBT! YOU! YES, YOU! CAN LEARN FROM HIM; SO THAT YOU KNOW WITH ABSOLUTE CERTAINTY WHAT IS TRUE! (and thereby resist any and all lies from anyone else and so never be deceived again.)

I'm asking/commanding (passing on God's Commandment) those who know Him already worldwide, to please focus on the Great Commission of our Lord Jesus Christ, the One True God, to tell everyone, your neighbors, and as many as possible who you meet in

life, that everyone needs to KNOW THE LORD! (Jer 31:34; Mt 28:18-20) Every day that goes by is another day of souls perishing in their sins for those who refuse to know the Living God, and every moment goes by that the saints are telling the rest of humanity of their immediate and dire need to Repent and Obey the Living God, by Getting Baptized in His Name and filled with Holy Spirit of Truth, is another day of suffering and death on earth. (evil continues and is perpetuated by everyone who does not know and is not obeying our Eternal Creator, Lord and Savior, Jesus Christ and is why in the end He casts all such unrepentant wicked persons into the lake of fire. Rev. 21:8)

Lies are propagated by the lying devil, who by his spirit of err/deception, has taken control of so many, (when you believe lies, you are deceived, by: 1) the Deceiving, Lying devil 2) by other persons who have been deceived 3) by your own delusions. THE ONLY WAY TO BE FREE FROM LIES OF ALL KINDS IS TO KNOW THE ONE WHO IS TRUTH AND THE ULTIMATE DEFINER OF WHAT IS TRUE IN ALL CREATION - GOD! JESUS CHRIST! (the English Name of the One True God - John 14:6-9; 8:32-36; 10:32-36; Acts 2:38-39; 4:12; 1Jn 5:19-21; 2Tim 2:26; 1Jn 4:1-6)

(Again, read that **The Holy Bible is Provably and Verifiably Divinely Inspired** and that in its contents "Jesus Christ" is the english Name for the One True God.)

SO, FIRST AND FOREMOST, the theory of evolution is not just false but IMPOSSIBLE, because God Almighty Himself says so! (Very first chapter of Genesis; the very first chapter God has publicly given mankind in the Holy Bible. In other words, RIGHT AWAY everyone should KNOW BEYOND ALL DOUBT, that the theory of evolution is false and impossible JUST BY READING THE FIRST CHAPTER of HIS WORDS in the Holy Bible! People everywhere MUST GRASP THIS FACT! The Holy Bible is the literary standard, the Foundation of Truth, God has given mankind by which ALL other opinions PAST, PRESENT and FUTURE are tried and tested by (not the other way around)! The moment ANYONE has an imagination contrary to those THOROUGHLY researched and WELL VERIFIED contents, is the moment they need to toss that vain imagination figuratively in the trash and never muse on it again! (like darwin's theory of evolution)

Since far too many today have factually been deceived (brainwashed), they actually laugh when I make such plain statements to them and won't even bother to listen, read or learn anything true for the first time in their existence. It's a very sad thing to see people who imagine the crap in their heads is actually knowledge, when it's all just a bunch of lies. I know how it feels to find out that your entire education regarding the origin of the universe and life is completely untrue, because you spent years of your life wrongly believing it was. Yes, I know, because I endured the same brainwashing everyone else currently is going through in coerced indoctrination today. It's a crying shame that the men and women who founded our nation and public education let antichrists rip it away from them. Rockefeller and other wealthy persons, with their sinister, greedy agendas took public education away from the people, the Church of Jesus Christ, because they wanted to use it to make slaves to their evil wills, instead of the REAL PURPOSE of public education, which is to teach everyone FROM THE CONTENTS OF THE HOLY BIBLE, Truth concerning the most important facts of history and reality! HOW TO KNOW OUR ETERNAL CREATOR, LORD and SAVIOR, JESUS CHRIST, personally!

YES! I SAID IT! PUBLIC EDUCATION CAME INTO EXISTENCE BECAUSE THE ONE TRUE GOD IN THE HOLY BIBLE COMMANDED US TO TEACH CHILDREN HIS COMMANDMENTS AND PRECEPTS FROM GENERATION TO GENERATION THEREIN! It was the Words of God in the Holy Bible, that caused many thousands of ancient scribes to write down and proclaim its contents to the masses throughout history. (The Holy Bible has tens of thousands of ancient manuscripts proving its veracity over time like no other book in all the world!) So within the Church of Jesus Christ on earth, monasteries of Biblical Scholars came into existence because God commands us to STUDY His Words! (scholarship on earth arose because God is REAL and told people to STUDY, LEARN, and treasure Wisdom and Knowledge!) that's why virtually all universities in western civilization have mottos attributed to that fact! GOD AND HIS WORDS ARE THE SOURCE OF WISDOM AND KNOWLEDGE AND THE VERY REASON PUBLIC EDUCATION EXISTS!

It's past time for Americans and nations of the world to properly put the Holy Bible back in our schools and universities as the primary text book in all years of education lifelong! NEVER again let devils remove it for any reason! EDUCATION EXISTS TO TEACH THE MOST IMPORTANT TRUTHS TO CHILDREN EVERYWHERE AND THAT IS EVERYONE NEEDS TO KNOW AND LEARN FROM THE ONE TRUE GOD, JESUS CHRIST, LIFELONG, NO MATTER WHAT THEY WISH TO SPECIALIZE IN OR DO WITH THEIR LIVES! GOD IS THE ONE WHO TEACHES ONLY TRUTH AND NEVER LIES! GOD IS THE ONE WHO KNOWS EVERYTHING! NO ONE KNOWS MORE THAN HE DOES! NO ONE! SO, HE IS THE TRUE TEACHER FOR US ALL! (again, no ifs, ands or buts!) GOD AND GOD ALONE IS THE ULTIMATE AUTHORITY! THE TEACHER OF teachers! NO ONE ELSE!

Stated as simply as I can state it, if you don't know Jesus Christ, who is the Truth and the Definer of All Truth, and aren't learning from Him personally, you are presently lacking the most important information on earth and are currently deceived. (are believing lies or one kind or another). There is no Greater Authority, no Greater Teacher, than God, Jesus Christ.

Founders of these United States of America like John Witherspoon stated:

"That he is the best friend to American liberty, who is most sincere and active in promoting true and undefiled religion, and who sets himself with the greatest firmness to bear down profanity and immorality of every kind. Whoever is an avowed enemy to God, I scruple not to call him an enemy to his country." - https://oll.libertyfund.org/pages/1776-witherspoon-dominion-of-providence-over-the-passions-of-men-sermon

So, if the Brave Patriots that fought for our freedoms and rights to the death were still with us today, they would look at any and all persons removing the Holy Bible and Truth (Creation Science) from public education today as absolute traitors (and I agree wholeheartedly). Our nation is going downhill FAST since the Holy Bible and prayer was removed from our public schools, solely because evil people like Rockefeller wanted to enslave Americans to their own wicked agendas; it's a crying shame people actually argue today that they a descendant of some ape in their past, rather than the Divine Creation, Son or Daughter of God, they were Created and Made by Him to be:

https://whatyouknowmightnotbeso.com/graphs.html - everything that was once commendable about our nation is in decline and what is deplorable is on the rise, since we started teaching lies to children in our schools instead of the truth that schools were originally founded to teach, the contents of the Holy Bible, the Words of God Almighty! Suddenly we have people openly marching in our streets waving abstracts of the Rainbow, the Sign God gave us that He wouldn't destroy us by a global flood again, in His Face and in the faces of all who know and love Him. Mocking the Living God, not realizing that He told us that if mankind dares to become so wicked again, He's going to incinerate them even worse than the destruction of Sodom and Gomorrah! So, the presence of such rebellion and wickedness on display in our nation threatens to leave such wicked cities in rubble, ruins and smoldering ashes! (as well as any nations that are tolerating such demonic madness to be on display publicly.)

TAKE A GOOD LOOK AMERICA! AND NATIONS OF THE WORLD! IS THIS WHAT YOU WANT TO HAPPEN TO YOU AND YOUR PEOPLE?! - https://www.youtube.com/watch?v=fHQEcqwrYxc ANY AND ALL PERSONS REJECTING THE ONE TRUE GOD AND HIS WORDS IN THE HOLY BIBLE ARE HEADED FOR HIS WRATH! (no ifs, ands or buts!) And as an American, I don't want that to happen to our nation!

http://www.blastthetrumpet.org/PublicLetters/AAAUpdatedPublicAlertsMattersofLifeandDeath/Updates053016/Americans%20and%20Citizens%20of%20the%20World%20FREEDOM%20and%20LIFE%20Comes%20from%20GOD.pdf - America was founded to be a Christian nation with the rights and freedoms afforded us thereby. Our liberties and freedom were for a moral citizenry. Letting amoral values corrupt our nation spells doom.

So, any Americans disdaining the contents of the Holy Bible are not just unpatriotic, they're subversives, and if they're in positions of power and influence like politicians, judges, teachers and professors, TRAITORS! Understand our nation, its laws, its public education system, public schools and universities are all the result of God, the Holy Bible, and those brave souls that fought to the death to found these United States!

Everyone needs to comprehend ASAP that virtually none of the crap regarding origin of the universe and life on earth being wrongly taught in the name of science today is actually science! It's all just imaginary hogwash of the worst kind! IT DOESN'T EVEN RISE TO THE MERIT OF DECENT SCIENCE FICTION! So, any public schools or judges supporting that crap and censoring Biblical Creation Science instead ARE TRAITORS! And yes, I'm publicly stating the reasons for these conclusions!

The well-founded, thoroughly established history of the Biblical Text should NEVER talk a back seat to fanciful notions about prehistory! Especially when they're wrong! Never once has the Holy Bible been disproven! It has only been proven, the more we know about creation the more it is proven as authentic and true Words of God Almighty.

Schools and universities have been taken over by antichrists, people that are ignorant, deceived and self-deluded, who don't know God, who don't know our history, who don't know much of anything truthful at all, and those who do, Biblical Creationists, have been getting fired and censored all by corrupt antichristian judges ruling in favor of those who imagine they're descended from monkeys and pond scum! Seriously, NONE OF THE EVOLUTIONARY PARADIGM regarding the origin of the universe and life is true! It's all a mountainous pile of imaginary, stinking crap, hatched up by God-denying fools! NONE OF IT IS TRUE! It's just a wild bunch of IMAGINATIONS that don't stack up to anything of merit.

https://www.facebook.com/photo/?fbid=6382847548460817&set=a.115635768515391 - YES! I DO UNDERSTAND THEIR STORIES! it's the reason I'm emphatically stating it's all totally imaginary NONSENSE! proposed by people DESPERATE to do away with the One True God and His Words in the Holy Bible. (most of those I've encountered doing so refuse to acknowledge even the possibility of God, especially the God of the Holy Bible, Jesus Christ, because they are factually evil people who are refusing to obey His Commandment to Repent of their wicked ways: sexually immorality is a very common trait among all God-deniers today: pedophiles, homosexuals, adulterers, fornicators, etc. etc.) In other words, they don't want to obey the Living God and so are deliberately not only corrupting themselves with lies in denying Him and His Words in the Holy Bible, but are evilly corrupting the minds of innocent children in our schools as well.

https://www.youtube.com/results?search_query=yuri+bezmenov+1983+interview+ - enemies of God and America, know that nations that fall away from God, suffer and can even be destroyed completely; so, infiltrators have indeed been deliberately demoralizing Americans in our schools, universities and through the media. Most of the corruption going on today is due to personal greed and vices like their own sexual immorality; so, the removal of the Holy Bible and prayer from our public schools has nothing whatsoever to do with science and everything to do with the intentional actions of very wicked, very evil, very immoral people!

http://www.blastthetrumpet.org/PublicLetters/AAAUpdatedPublicAlertsMattersofLifeandDeath/Updates053016/The%20cult%20of%20modern%20scientism%20versus%20the%20Science%20of%20Biblical%20Creation.pdf

http://www.blastthetrumpet.org/PublicLetters/AAAUpdatedPublicAlertsMattersofLifeandDeath/Updates053016/THE%20PLANNED%20DEMORALIZATION%20OF%20AMERICA.pdf - demoralization isn't just the process of getting people to believe lies, it's the process of intentionally corrupting them, making them become wicked and depraved like those behind the process itself!

So, the "dummying down" of Americans and "perverting" Americans by infiltrating schools and media is deliberate and it takes Americans who are still decent, still moral and still in their right minds to stand up and so no to these overt attacks on them and their children.

https://www.youtube.com/results?search_query=sexualizing+children+in+our+schools - this evil is factually being funded by wealthy immoral persons who, like Jeffrey Epstein arrested for crimes against children, teach children lies like the theory of evolution to exploit them; it facilitates over-sexualizing children and so child-trafficking exists around the world, (no matter what the scandal is called and no matter how hard they try to deny them, there is a very real problem with sexually immoral people, even pedophiles, in positions of power and influence and a very real problem of sexual exploitation of children taking place in America and other nations globally) and pretty much all wicked child trafficking going on in this world. It is EVIL; people who are DELIBERATELY attacking children that are responsible and why when I even mention the Holy Bible to them, they erupt in blasphemy, mockeries, lies and laughter, while they refuse to learn anything true, because sadly, these children, young adults for several generations now, have all factually been demoralized, brainwashed and are a danger to themselves, our nation and the world as such. (Everyone believing lies, passes them on; becomes liars themselves and God tells us He is going to cast all liars into the lake of fire. Rev 21:8) Furthermore, He tells us that all wicked rebels still mocking, blaspheming, still sexually immoral, at His

Coming are going to be incinerated in His Wrath. (2Th 1:8-9) and end up in the lake of fire. Rev 21:8

In addition, nations like America that was founded on His Precepts in the Holy Bible, that depart from Him suffer growing Divine Consequences, until the people are suffering and dying and the nation falls! Deuteronomy 28 - Divine Curses/Consequences are seen throughout history to this very day, anyone denying them is ignoring the most important facts of history and reality before our eyes.

IF YOU HAVE GOTTEN THIS FAR, YOU NOW UNDERSTAND WHY THE NUMBER ONE REASON EVOLUTION IS FALSE AND IMPOSSIBLE IS BECAUSE THE ONE TRUE GOD PLAINLY SAYS SO IN HIS WORDS GIVEN TO US PUBLICLY, THE CONTENTS OF THE HOLY BIBLE. IF YOU READ THE OTHER BOOKS CITED ON THE SUBJECT YOU NOW COMPREHEND JUST SOME OF THE REASONS WHY THE HOLY BIBLE IS TRUE, PROVABLY AND VERIFIABLY DIVINELY INSPIRED AND YES, IS TRUE ABOUT SCIENCE! "SCIENCE" IS JUST ANOTHER WORD FOR "KNOWLEDGE"! The "scientific method" is just the methods we use to acquire knowledge! So, HISTORY is Science! and it's a large area of Science! Most all teaching lies in the name of science today are particularly ignoring HISTORY! And the most important facts of history are contained in the Holy

Bible. **SO YES! THE HOLY BIBLE IS MOST DEFINITELY A BOOK OF SCIENCE!**

https://www.facebook.com/photo.php?fbid=7203384469740450&set=pb.100002069048072.-2207520000&type=3 - The contents of the Holy Bible are factually the reasons other books even exist in western civilization. It was those contents that inspired the Gutenberg Printing Press and with the invention of movable type can the advent of modern books. With the advent of modern books, came the founders of the fields of the modern sciences. The Holy Bible has been studied globally for centuries under all those fields of science BECAUSE God says something about all those fields of science in its contents!

http://www.blastthetrumpet.org/PublicLetters/AAAUpdatedPublicAlertsMattersofLifeandDeath/Updates053016/Fathers%20of%20the%20Sciences%20and%20Great%20Scientific%20Minds.pdf

Contrary to the opinion of far too many these days, science is in no opposition to God our Creator or the Divinely Inspired Biblical Account. In fact, science is the result of the inquisitive nature of mankind as he perceives all creation. The fathers of the many fields of science today mostly were just persons who sought to know what is

Truth. You will find in their own words that they not only believed in the Biblical Account; but that they believe in the GOD, our Creator of the Biblical Account. As you read these quotes, understand their point and mine; and that is they sought to know more about our Creator by the study of His Creation; as should we all.

Sir Francis Bacon, known as the father of the scientific method, wrote: **"There are two books laid before us to study, to prevent our falling into error; first, the volume of Scriptures, which reveal the will of God, then the volume of the Creatures [Creation], which express His power."** - http://www.ucg.org/booklet/bible-true/one-worlds-most-popular-books/their-own-words-great-men-and-women-who-highly-resp/

http://www.brainpickings.org/index.php/2013/02/15/galileo-letter-to-duchess-of-tuscany/ - contrary to the title of this blog, **Galileo Galilei** was in no way opposing GOD and he said this plainly in his own words.

"I do not feel obliged to believe that the same God who has endowed us with senses, reason and intellect has intended us to forego their use and by some other means to give us knowledge which we can attain by them. He would not require us to deny sense and

reason in physical matters which are set before our eyes and minds by direct experience or necessary demonstrations."

and

"People who are unable to understand perfectly both the Bible and the science far outnumber those who do understand them. The former, glancing superficially through the Bible, would arrogate to themselves the authority to decree upon every question of physics on the strength of some word which they have misunderstood, and which was employed by the sacred authors for some different purpose. And the smaller number of understanding men could not dam up the furious torrent of such people, who would gain the majority of followers simply because it is much more pleasant to gain a reputation for wisdom without effort or study than to consume oneself tirelessly in the most laborious disciplines."

and he goes on to state

"...nor is God any less excellently revealed in Nature's actions than in the sacred statements of the Bible."

he was even familiar with and understood the chief endeavor of the Holy Ghost (the Holy Spirit).

"Now if the Holy Spirit has purposely neglected to teach us propositions of this sort as irrelevant to the highest goal (that is, to our salvation), how can anyone affirm that it is obligatory to take sides on them, that one belief is required by faith, while the other side is erroneous?"

Being filled with the Holy Spirit myself; if present; I would have replied to him, that GOD our Creator has not *neglected* to teach us Truth in any way; but that we will be learning it from Him for all eternity. Our present limited understanding of what is Truth; is not to be laid in any way upon the Master of the Universe; as if our Creator has not made it plain before us; as we adventure in discovery of it; by observing and meditating upon His Creation; including the diligent study of His Word given to us in the Holy Bible. Please do not sir, out of your own present ignorance, falsely accuse our Creator of some supposed negligence on His part; and I say the same to the many who are doing so today.

I will furthermore add, that if you THINK you have found something contrary to the Biblical Account in the study of Creation; then you should examine both what you THINK is contrary and your comprehension of what you read more thoroughly. Galileo did properly understand that most who argue from what is contained therein against fundamental, observable Truth; do so because they have incorrectly interpreted the meaning of what they read; even as he did in presupposing others before him had not observed such things as he had through his optics. (http://www.godempowersyou.com/documentation/HistoricalTestimonyandMinistry/MansChronologicalExistenceExplained.pdf) For contrary to what Galileo went on to state; (for which I would have rebuked him to his face sternly):

Galileo stated falsely; **"That the intention of the Holy Ghost is to teach us how one goes to heaven, not how heaven goes."**

Our Creator teaches us only Truth about **EVERYTHING**! As it is written:

"and the Comforter, the Holy Spirit, whom the Father will send in my name, he will teach you all things, and remind you of all things that I said to you." - http://biblehub.com/john/14-26.htm

"But whenever The Spirit of The Truth comes, he will lead you into the whole truth, for he will not speak of his own will, but he shall speak whatever he shall hear and he shall reveal the future to you." - http://biblehub.com/john/16-13.htm

Our Eternal Creator teaches us the Truth concerning everything! Lack of knowledge is apparent in us all because we are not our Perfect Eternal Creator; but it is most apparent in those who are choosing not to learn from Him directly (by the Presence of His Holy Spirit upon us http://www.biblegateway.com/passage/?search=Acts+2%3A38-39&version=KJV; diligent study of His Word http://biblehub.com/2_timothy/2-15.htm, the Holy Bible, and the confirming comparative, accurate observation of His Creation; the sciences http://biblehub.com/romans/12-2.htm)!

http://biblehub.com/1_john/2-27.htm and http://biblehub.com/hebrews/6-18.htm

It seems some people like to argue just for the sake of arguing. Heliocentric! Geocentric! Heliocentric! Geocentric! etc. but the Truth is that our entire solar system is in constant motion hurling through space like a complex bola. That BOTH the earth is revolving around the sun; and our solar system which is hurling around our galaxy; which is hurling through space itself in similar fashion.

http://www.youtube.com/watch?v=RON9rWcALWc

Even if one believes in the nonsense of the big bang, the universe had an origin. Their imaginary version of a singularity, would be the origin, why has no one tried to find it?

I tend to think what God showed me of how He could create a new universe best explains this one. He simply created all matter and energy into one huge superheated ENORMOUS BURNING MASS and then put his hands around it all and sent it spinning. Bits of matter and energy went spinning off, some exploding, and so on to form the galaxies. Some collided and formed the galaxies that are atypical or not all spinning in the same direction, but generally speaking the universe is simply expanding due to common laws of motion from when God himself created it all. GOD FORMED IT ALL NEW, but then tells us this creation is subject to decay, not evolving by any stretch of the imagination! It is all wearing out, growing old. So no, bits of matter and energy did not create themselves, chemicals neither, the stars and galaxies did not form the way anyone who believes what science is teaching these days regarding the expansion of space and time from the big bang. All that is imaginary nonsense, GOD CREATED IT ALL - all matter and energy, chemicals, stars and planets, galaxies - they were all made by Him in

the Beginning. The entire universe is like a massively complex bola of galaxies (more bolas) of solar systems (more bolas) of planetary systems (more bolas), etc.; that was sent spinning from the origin (the only stationary point in all time and space) and can be found at the intersection of time and space by reverse extrapolation of the expansion of the flung galaxies all flying outward from the origin.
(http://www.youtube.com/watch?v=ri3R6vFVfEg and http://www.youtube.com/watch?v=u-xUAC9ya3o and http://www.facebook.com/photo.php?fbid=451991028213195&set=a.115635768515391.22520.100002069048072&type=3&theater) our Creator set the entire universe in motion and in a way that galaxies are revolving around the origin while revolving around each other as they expand outward in the void of space. In the galaxies, solar systems are doing the very same thing; including ours. These solar systems are in motion in such a way as while they are revolving around their respective galactic centers, they have smaller/larger solar systems all around them do the very same thing. Universal Time is relative to the distance traveled from the origin since the Beginning. God tells us to measure time, seasons, years, by the movements of the stars, the most prominent heavenly bodies as seen from earth, the sun and moon. So, the passage of time is relative to the movements of the stars. The best analogy I could think of at present is the flung bola; while the sun is larger than the earth both are still in fact moving through space and revolving around each other as they do. The path through space is not such a

smooth line as is often depicted in diagrams but has the same kind of wobbles to it that flung bolas do (to a greater or lesser extent). So actually the paths are not like the 3d models in grade school at all but more like bolas and if you were to follow the paths would look like spiral waves as these particles of dust and fire of various sizes expand explosively from the origin of the universe (the same way light is observed to travel - http://www.howstuffworks.com/light6.htm - because when we look into time and space (all that we perceive of visible creation) we are actually viewing those spiral waves of light trails; not the moving objects themselves). SOME visible elements are spinning in seeming opposite directions from the vast majority BECAUSE OF EXPLOSIONS AND COLLISIONS WITH OTHER SPINNING BITS OF MATTER in space that we call galaxies today. Likewise, SOME particles don't appear in the many clusters of galaxies because of EXPLOSIONS OF SUPERHEATED matter we call stars, novas, etc. SOME of those particulates then cool and even ice and we call those objects things like dark matter, planets, asteroids, comets, etc. BUT GENERALLY, in 360 degrees from the origin in great orbital (wobbling) planes around the origin galaxies are spinning and revolving at various velocities (DUE TO THE AMOUNT OF MATTER THAT BROKE OFF FROM THE ORIGIN WHEN THE UNIVERSE WAS ALL GATHERED TOGETHER IN ONE SUPER HEATED ENORMOUS MASS AND IF IN THE PROCESS IT BROKE AWAY SLOWLY (LOW VISCOSITY GALAXY) OR QUICKLY (HIGH VISCOSITY GALAXY) OR EVEN EXPLODED FROM THE ORIGIN. To this day stars can and do explode (YES,

BURNING LIQUIDS, GASES AND SOLIDS CAN DO THAT) and even galactic centers (JUST LARGER MASSES THAN THE STARS OF SOLAR SYSTEMS REVOLVING AROUND THEM) can do the same.

My point here is that most of us are like little children (even Galileo) that when we explore the universe in which we find ourselves and discover anything at all about it are likely to boast of it to others and in such a way as to become very defensive about our "new discovery" (http://biblehub.com/ecclesiastes/1-9.htm); so defensive we can fall into the same heresy we accuse others of practicing (that of ignorance; made worse by the mixture of arrogance in far too many). And in that boast make errors like thinking GOD doesn't mean what He states in His Word; when in fact it is appropriate that in some regards it talks of the heavens from a geocentric perspective; because that is where we live in flesh and blood and how we look out and perceive the celestial bodies. There is therefore no error in that perspective and no contradiction, scientifically speaking or in any other rational connotation, throughout the Holy Bible. If one takes a larger rock of more mass and ties it to a smaller and throws it in bola fashion both rocks still act upon one another to a greater and lesser extent; our solar system, while having more objects than just the two, operates in similar fashion; it is therefore neither heliocentric or geocentric, technically speaking. The reason some might think of the Holy Bible as geocentric is simply God placed us here and we observe the heavens from here, but the Bible is not just telling us creation from our perspective but God is telling us creation from His Perspective as well. It is about our Creator dealing with mankind over its generations. http://biblehub.com/colossians/1-17.htm.

"As a man who has devoted his whole life to the most clear headed science, to the study of matter, I can tell you as a result of my research about atoms this much: There is no matter as such. All matter originates and exists only by virtue of a force which brings the particle of an atom to vibration and holds this most minute solar system of the atom together. We must assume behind this force the existence of a conscious and intelligent mind. This mind is the matrix of all matter." - -
Max Planck
(founder of the quantum theory and one of the most important physicists of the twentieth century)

http://www.simpletoremember.com/articles/a/science-quotes/

The mass of our sun (and all stars) is burning off; if it loses its mass at a rate that it's gravitational force equals the centripetal force of those objects flying around it and whose velocities are slowing down, then harmony is maintained; but in the more likely event either velocities of planets change to upset that balance (most likely by collisions or near collisions) or the mass of the star(sun) explodes or burns off; the earth could find itself either flung off into space or pulled into the sun's burning flames. It is; after all, only a lake of fire. http://biblehub.com/revelation/20-14.htm (I am NOT saying the sun is THE lake of fire); I am only saying it is A lake of fire (one of many billions of them) to make my point in how many people mock the Divinely Inspired Holy Scriptures while in the presence of such clear examples that such things as "lakes of fire" exist all around them; throughout the entire universe, and one very prominently viewed and experienced by all who dwell on earth day after day!

Albert Einstein

"In view of such harmony in the cosmos which I, with my limited human mind, am able to recognize, there are yet people who say there is no God. But what really makes me angry is that they quote me for the support of such views." - http://www.einsteinandreligion.com/atheism.html

While what we may have of Einstein's quotes may not categorically place him easily in any labeled worldview; it is obvious He believed in a Creator; after giving thought to what he observed in creation. (and I hope for his sake he determined who that "Spirit" is; before he breathed his last) http://www.clockbackward.com/2009/02/08/was-albert-einstein-religious/

"I want to know God's thoughts; the rest is just details." - http://jick.net/~jess/p200/quotes.html

Too many today are treating science and scientists as if they are infallible and speak almost with reverence about science and scientists. Most doing so seem not to understand that scientists (and science thereby) are fallible; even Einstein: https://www.blastthetrumpet.org/PublicLetters/AAAUpdatedPublicAlertsMattersofLifeandDeath/EequalsmisRubbish.pdf (much of science today just have cultic believers with no substantiation for those believes. EINSTEIN'S THEORY HAS BEEN DISPROVEN; by quantum mechanics, spooky action, and so on. TIME IS NOT A PHYSICAL PROPERTY - can anyone hand me time? So therefore, physical energy cannot be equated/relative to TIME! Too many championing science act as if they don't realize that even words and their definitions can change due to subjective opinions of people (including scientists) concerning them; let alone "scientific laws" or "scientific theories" http://en.wikipedia.org/wiki/Superseded_scientific_theories And obviously, those revering fallible men (science and scientists) above the Divinely Inspired Record of God our

Creator in the Holy Bible; don't realize that science and scientific discoveries are STILL trying to catch up with the knowledge in it! (knowledge that existed centuries or millenniums BEFORE scientists finally figure out the Truth of its contents): http://www.ucg.org/booklet/bible-true/bible-and-science/ It took the advent of supercomputers to find that there are prophetic encrypted messages within the Holy Bible placed there while it accurately recorded history and truth about creation simultaneously! https://www.google.com/search?q=the+bible+code+michael+drosnin&rlz=1C1GIGM_enUS535US535&oq=the+bible+code&aqs=chrome.2.69i57j0l5.8780j0j8&sourceid=chrome&espv=210&es_sm=93&ie=UTF-8 Very specific prophecies even with dates included in numerous instances!

Back to my point, science and scientists who accurately study creation will only find it leads to the Creator; no matter what field of science one may specialize in. Science will only continually prove the validity of the contents of the Holy Bible as it is still acquiring the knowledge it contains so plainly in the Divinely Inspired words therein: https://www.blastthetrumpet.org/PublicLetters/AAAUpdatedPublicAlertsMattersofLifeandDeath/Updates053016/NO%20MORE%20BRAINWASHING%20p1.pdf and https://www.blastthetrumpet.org/PublicLetters/AAAUpda

tedPublicAlertsMattersofLifeandDeath/Updates053016/NO%20MORE%20BRAINWASHING%20p2.pdf and https://www.blastthetrumpet.org/PublicLetters/AAAUpdatedPublicAlertsMattersofLifeandDeath/Prophecies%20Show%20Divine%20Inspiration%20of%20the%20Holy%20Bibl2.pdf And all these fathers of the sciences understood that.

http://en.wikipedia.org/wiki/List_of_Christian_thinkers_in_science

Max Planck - **"Both religion and science need for their activities the belief in God, and moreover God stands for the former in the beginning, and for the latter at the end of the whole thinking. For the former, God represents the basis, for the latter – the crown of any reasoning concerning the world-view." -** http://withalliamgod.wordpress.com/2010/11/28/max-planck-on-god/

Carl Sagan

"Science is not only compatible with spirituality; it is a profound source of spirituality. When we recognize our place in an immensity of light-years and in the passage of ages, when we grasp the intricacy, beauty,

and subtlety of life, then that soaring feeling, that sense of elation and humility combined, is surely spiritual...The notion that science and spirituality are somehow mutually exclusive does a disservice to both." - http://www.youtube.com/watch?v=Ag6fH8cU-MU As a scientist you would think he could answer such simple questions he expressed in that video: https://www.blastthetrumpet.org/PublicLetters/AAAUpdatedPublicAlertsMattersofLifeandDeath/Updates053016/Acknowledging%20the%20Eternal%20Creator%20Takes%20No%20Faith%20It%20is%20Scientific%20Fact.pdf and http://www.facebook.com/notes/michael-swenson/creation-v-evolution-please-use-your-intelligence-to-recognize-intelligence/566154943463469

Francis Collins

"Science is...a powerful way, indeed - to study the natural world. Science is not particularly effective...in making commentary about the supernatural world. Both worlds, for me, are quite real and quite important. They are investigated in different ways. They coexist. They illuminate each other."

--American physician-geneticist and director of the National Human Genome Research Institute

-

http://www.goodreads.com/author/quotes/20100.Francis_S_Collins

William H. Bragg

"From religion comes a man's purpose; from science, his power to achieve it. Sometimes people ask if religion and science are not opposed to one another. They are: in the sense that the thumb and fingers of my hands are opposed to one another. It is an opposition by means of which anything can be grasped."

--British physicist, chemist, and mathematician. Awarded Nobel Prize in 1915

- http://www.beliefnet.com/Faiths/Galleries/God-and-Science-What-Do-Scientists-Believe.aspx?p=9

And I agree that the way to grasp the Truth contained in writing within the Holy Bible comes from seeking answers from our Creator and by observing Creation in a rational manner; scientifically. Even our Creator told us that any claims that scientific findings were in anyway contradictory to His Divine Instructions for Life contained in the Holy Bible were false; as true science (factual observations of creation) will always confirm what our Creator has declared and made known to us. No one who long observes and meditates upon what he or she perceives can fail to see the Intelligence that made all things and holds all things together to this very moment. http://biblehub.com/1_timothy/6-20.htm

Fred Hoyle

"A junkyard contains all the bits and pieces of a Boeing 747, dismembered and in disarray. A whirlwind happens to blow through the yard. What is the chance that after its passage a fully assembled 747, ready to fly, will be found standing there? So small as to be negligible, even if a tornado were to blow through enough junkyards to fill the whole Universe."

"Life cannot have had a random beginning ... The trouble is that there are about 2000 enzymes, and the chance of obtaining them all in a random trial is only one part in 10^40,000, an outrageously small probability that could not be faced even if the whole universe consisted of organic soup."

"The notion that not only the biopolymer but the operating program of a living cell could be arrived at by chance in a primordial organic soup here on the Earth is evidently nonsense of a high order."

"Once we see, however, that the probability of life originating at random is so utterly miniscule as to make it absurd, it becomes sensible to think that the favorable properties of physics on which life depends are in every respect deliberate .... It is therefore almost

inevitable that our own measure of intelligence must reflect ... higher intelligences ... even to the limit of God ... such a theory is so obvious that one wonders why it is not widely accepted as being self-evident. The reasons are psychological rather than scientific."

B. Prokop said...

Add Fred Hoyle. British cosmologist; came up with the term "The Big Bang".

Fierce and vocal atheist for decades. Embraced theism late in life, after being convinced that Intelligent Design was the only rational belief.

(Also a great writer. Might I recommend "The Black Cloud".) -
http://dangerousidea.blogspot.com/2011/03/list-of-former-atheists.html
--English mathematician and astronomer.

-

http://www.goodreads.com/author/quotes/199992.Fred_Hoyle

I'm also a former atheist (the brainwashing in our American public schools is that effective; so effective I forgot about early childhood recurring nightly dreams from our Creator for many years).

Johannes Kepler

1571-1630 German Astronomer

(planetary orbits, optics, mathematical language of science)
The 3 laws of planetary motion. Advanced Copernicus' heliocentric theory.
"Since we astronomers are priests of the highest God in regard to the book of nature, it befits us to be thoughtful, not of the glory of our minds, but rather, above all else, of the glory of God."

"...the harmony in the universe is "a sacred sermon, a veritable hymn to God the Creator".... "Oh God, I am thinking thy thoughts after Thee" Johannes Kepler

"I believe only and alone in the service of Jesus Christ. In him is all refuge and solace."

http://www.valleypresbyterian.org/curriculum/science/quotes.htm

Isaac Newton

English mathematician & physicist, one of the greatest scientists of all time.
Laws of gravitation and motion, developed calculus. Major contributions to optics, physics, math and astronomy.

“He who thinks half-heartedly will not believe in God; but he who really thinks has to believe in God.”

The solar system itself could not have been produced by blind chance or fortuitous causes but only by a cause "very well skilled in mechanics and geometry."

"There are more sure marks of authenticity in the Bible than in any profane history."

"No sciences are better attested to than the science of the Bible." "This most beautiful system of the sun, planets, and comets, could only proceed from the counsel and dominion of an intelligent and powerful Being.... This Being governs all things, not as the soul of the world, but as Lord over all; and on account of his dominion he is wont to be called Lord God." Isaac Newton; Principles.

"It must be expressed in the very form of sound words in which it was delivered by the apostles. For men are apt to run into partings about deductions. All the old heresies lie in deductions. The true faith was in the Biblical texts."

http://www.valleypresbyterian.org/curriculum/science/quotes.htm

“This most beautiful system of the sun, planets and comets, could only proceed from the counsel and dominion of an intelligent and powerful Being.”

http://www.simpletoremember.com/articles/a/science-quotes/

James Clerk Maxwell

Scottish Physicist

discovered the relationships between electricity, magnetism and light and introduced the concept of field of EM force, devout Christian

"Happy is the man who can recognize in the work of today a connected portion of the work of life, and an embodiment of the work of Eternity."

"Think what God has determined to do to all those who submit themselves to his righteousness and are willing to receive his gift [of eternal life in Jesus Christ]. They are to be conformed to the image of his Son and when that is fulfilled and God sees they are conformed to the image of Christ, there can be no more condemnation."

"I think that men of science as well as other men need to learn from Christ, and I think that Christians whose minds are scientific are bound to study science that this view of the glory of God may be as extensive as their being is capable of."

"The more we enter into Christ's work He will have more room to work His work in us. For He always

desires us to be one with us. Our worship is social, and Christ will be where two or three are gathered together in His name."

http://www.adherents.com/people/pm/James_Clerk_Maxwell.html

http://www.valleypresbyterian.org/curriculum/science/quotes.htm

Michael Faraday

Discovered benzene and electromagnetic radiation, invented the generator and was the main architect of classical field theory. One of the prominent experimental scientists of all time.

"Speculations, man, I have none. I have certainties. I thank God that I don't rest my dying head upon speculations for "I know whom I have believed and am persuaded that he is able to keep that which I've committed unto him against that day."

http://www.valleypresbyterian.org/curriculum/science/quotes.htm

http://www.icr.org/article/3958/

http://www.answersingenesis.org/articles/cm/v12/n4/faraday

Robert Boyle

Developed the idea of atoms. Boyle's law.

Wrote: *The Wisdom of God Manifested in the Works of Creation*. Governor of the Corporation for the Spread of the Gospel of Jesus Christ in New England.

"From a knowledge of God's work we shall know Him."

http://www.goodreads.com/quotes/144147-from-a-knowledge-of-god-s-work-we-shall-know-him

Arthur Compton

Nobel Prize-winning physicist for his discovery and explanation of the change in the wavelength of X rays when they collide with electrons. Compton effect confirmed the dual nature of electromagnetic radiation as both a wave and a particle.

"From earliest childhood I have learned to see in Jesus the supreme example of one who loves his neighbors and expresses that love in actions that count, who knows that people can find their souls by losing themselves in something of great value, who will die rather than deny the truth in favor of the popular view held by his most respected contemporaries. That Jesus lives so vitally in men today makes me hope that by

following in his footsteps in my small way I also may live forever."

"For myself, faith begins with the realization that a supreme intelligence brought the universe into being and created man. It is not difficult for me to have this faith, for it is incontrovertible that where there is a plan there is intelligence. An orderly, unfolding universe testifies to the truth of the most majestic statement ever uttered: 'In the beginning God ... .'"

"In their essence there can be no conflict between science and religion. Science is a reliable method of finding truth. Religion is the search for a satisfying basis for life."

"What nobler ambition can one have than to cooperate with his Maker in bringing about a better world in which we can live? Science has created a world in which Christianity is a necessity."

"I believe that its insistence on the inherent value of individual men and women Christianity has the key to survival and the good life in the modern world."

http://www.doesgodexist.org/JulAug08/Nobel-Compton.html

J.J. Thompson

Discovered the electron

"In the distance tower still higher [scientific] peaks which will yield to those who ascend them still wider prospects and deepen the feeling whose truth is emphasized by every advance in science, that great are the works of the Lord." (Statement in *Nature*)

http://www.valleypresbyterian.org/curriculum/science/quotes.htm

Charles Coulson

Architect of molecular orbital theory.

"There were some ten of us and together we sought for God and together we found Him. I learned for the first time in my life that God was my friend. God became real to me, utterly real. I knew Him and could talk with Him as I never imagined it before and these prayers were the most glorious moment of the day. Life had a purpose and that purpose colored everything."

http://www.valleypresbyterian.org/curriculum/science/quotes.htm

Robert Griffith, a member of our U.S. Academy of Sciences, Otto Stern professor of physics at Carnegie Mellon University received one of the most coveted awards of the American Physical Society in 1984 on his work in physical mechanics and thermodynamics. Physics

Today said he is an evangelical Christian who is an amateur theologian and who helps teach a course on Christianity and science.

He recently said:

"If we need an atheist for a debate, I'd go to the philosophy department—the physics department isn't much use."

At Berkeley University, among 55 chemistry professors, we only had one who was willing to openly identify himself as an atheist, my good friend Bob, with whom I still have many discussions about spiritual things.

http://bib.irr.org/scientists-and-their-gods

Charles H. Townes

Nobel Prize winner. Inventor of the maser and laser. First to observe an interstellar molecule

"You may well ask, "Where does God come into this," to me, that's almost a pointless question. If you believe in God at all, there is no particular "where"—He is always there, everywhere....To me, God is personal yet omnipresent. A great source of strength, He has made an enormous difference to me." From *Making Waves*.

http://www.valleypresbyterian.org/curriculum/science/quotes.htm

"For successful science of the type we know, we must have faith that the universe is governed by reliable laws and, further, that these laws can be discovered by human inquiry. The logic of human inquiry is trustworthy only if nature is itself logical.... This is the faith of reason"

http://benevolentbaptist.wordpress.com/2013/02/02/faith-within-reason-charles-h-townes-on-faith-in-scientific-thought/

Allan Sandage

One of the world's greatest observational cosmologists. Estimated the age of the universe.

"The world is too complicated in all its parts and interconnections to be due to chance...I am convinced that the existence of life with all its order and each of its organisms is simply too well put together."

http://www.valleypresbyterian.org/curriculum/science/quotes.htm

"I find it quite improbable that such order came out of chaos. There has to be some organizing principle. God to me is a mystery but is the explanation for the miracle of existence, why there is something instead of nothing."

- Alan Sandage (winner of the Crawford prize in astronomy) Willford, J.N. March 12, 1991. Sizing up the Cosmos: An Astronomers Quest. New York Times, p. B9.

http://www.simpletoremember.com/articles/a/science-quotes/

Ilya Prigogine (Chemist-Physicist)
Recipient of two Nobel Prizes in chemistry

"The statistical probability that organic structures and the most precisely harmonized reactions that typify living organisms would be generated by accident, is zero."

I. Prigogine, N. Gregair, A. Babbyabtz, Physics Today 25, pp. 23-28

http://www.simpletoremember.com/articles/a/science-quotes/

Christian de Duve

"A Guided Tour of the Living Cell" (Nobel laureate and organic chemist)

"If you equate the probability of the birth of a bacteria cell to chance assembly of its atoms, eternity will not suffice to produce one... Faced with the enormous sum of lucky draws behind the success of the

evolutionary game, one may legitimately wonder to what extent this success is actually written into the fabric of the universe."

http://www.simpletoremember.com/articles/a/science-quotes/

"[Probably the leading paleontologist alive today, Simon Conway Morris, the scientist who discovered the significance of the Cambrian explosion of animal life, writes in his seminal book, Life's Solutions, that he is "convinced" that nature's success in the lottery of life has "metaphysical implications."]"

http://www.simpletoremember.com/articles/a/science-quotes/

Dr. Michael Denton (Australian microbiologist)

"...The capacity of DNA to store information vastly exceeds that of any other known system: it is so efficient that all the information needed to specify an organism as complex as man weighs less than a few thousand millionths of a gram. The information necessary to specify the design of all the species of organisms which have ever existed on the planet...could be held in a teaspoon and there would still be room left for all the information in every book ever written..."

http://www.simpletoremember.com/articles/a/science-quotes/

John O'Keefe (astronomer at NASA)

"We are, by astronomical standards, a pampered, cosseted, cherished group of creatures.. .. If the Universe had not been made with the most exacting precision we could never have come into existence. It is my view that these circumstances indicate the universe was created for man to live in."

- Heeren, F. 1995. Show Me God. Wheeling, IL, Searchlight Publications, p. 200

http://www.simpletoremember.com/articles/a/science-quotes/

Anthony Flew

Professor of Philosophy, former atheist, author, and debater

"It is, for example, impossible for evolution to account for the fact that one single cell can carry more data than all the volumes of the Encyclopedia Britannica put together."

"It now seems to me that the findings of more than fifty years of DNA research have provided materials for a new and enormously powerful argument to design."

http://www.simpletoremember.com/articles/a/science-quotes/

George Greenstein

"As we survey all the evidence, the thought insistently arises that some supernatural agency—or, rather, Agency—must be involved. Is it possible that suddenly, without intending to, we have stumbled upon scientific proof of the existence of a Supreme Being? Was it God who stepped in and so providentially crafted the cosmos for our benefit?"

(American astronomer) *Greenstein, George. The Symbiotic, Universe: Life and Mind in the Cosmos. (New York: William Morrow, (1988), pp.* 26-27

http://www.simpletoremember.com/articles/a/science-quotes/

Frank Tipler

"When I began my career as a cosmologist some twenty years ago, I was a convinced atheist. I never in my wildest dreams imagined that one day I would be writing

a book purporting to show that the central claims of Judeo-Christian theology are in fact true, that these claims are straightforward deductions of the laws of physics as we now understand them. I have been forced into these conclusions by the inexorable logic of my own special branch of physics."

(Professor of Mathematical Physics) Tipler, F.J. 1994. The Physics Of Immortality. New York, Doubleday, Preface.

http://www.simpletoremember.com/articles/a/science-quotes/

William Phillips

Nobel prize for the development of methods to cool and trap atoms with laser light.

"God has given us an incredibly fascinating world to live in and explore."

Arthur Schawlow

Nobel Prize winner in Physics. Development of the laser and laser spectroscopy

"We are fortunate to have the Bible, and especially the New Testament, which tells so much about God in widely accessible, human terms."

Alexander Polyakov

Russian Physicist

"We know that nature is described by the best of all possible mathematics because God created it." The famous Russian physicist, Alexander Polyakov Fortune magazine (October, 1986),

Sir William Herschel

"All human discoveries seem to be made only for the purpose of confirming more and more the Truths contained in the Sacred Scriptures."

Sir William Herschel (1738-1822), English astronomer, he made numerous discoveries about the laws of the heavens.

C.F.Gauss

One of the greatest mathematicians of all time. Contributed significantly to pure mathematic and made practical applications of importance for 20th-century astronomy, geodesy, and electromagnetism.

"There are problems to whose solution I would attach an infinitely greater importance than to those of mathematics, for example touching ethics, or our relation

to God, or concerning our destiny and our future; but their solution lies wholly beyond us and completely outside the province of science."

John Wheeler

Princeton University astronomer

"Slight variations in physical laws such as gravity or electromagnetism would make life impossible . . . the necessity to produce life lies at the center of the universe's whole machinery and design," stated John Wheeler, Princeton University professor of physics *(Reader's Digest, Sept., 1986).*

http://www.valleypresbyterian.org/curriculum/science/quotes.htm

Lord Kelvin

1824-1907 (William Thomson) British physicist; helped lay the foundations of thermodynamics.
First and second laws of thermodynamics. Absolute temp scale. Trans-Atlantic cable.

"I believe that the more thoroughly science is studied, the further does it take us from anything comparable to atheism."

Do not be afraid to be free thinkers. If you think strongly enough, you will be forced by science to the belief in God.

http://www.valleypresbyterian.org/curriculum/science/quotes.htm and http://www.valleypresbyterian.org/curriculum/science/science.htm

Blaise Pascal, *Pensées*

"There is a God shaped vacuum in the heart of every man which cannot be filled by any created thing, but only by God, the Creator, made known through Jesus."

-

http://www.goodreads.com/author/quotes/10994.Blaise_Pascal

About 80% of the listed 100 most influential persons in history acknowledge the Creator as the God of the Holy Bible; most of the other 20% in this list were influential but in a detrimental way toward humanity and creation. http://www.adherents.com/adh_influ.html

It is evident that the greatest minds and most powerful persons in the history of the world have been deeply religious; purpose driven people. (even atheism is religious in nature in that they choose to believe in a concept that they can in no way prove and choose to believe in it; **even though science and rational thought <u>proves</u> their faith in the absence of an Intelligent, Eternal Creator is false -** <u>http://www.facebook.com/notes/michael-swenson/no-more-brainwashing-no-more-lies-science-proves-who-god-is-part-1/518695411542756</u>) But my point in this brief note on these scientific greats is that even the fields of science we have today would not even exist apart from those who wanted to know more about our Creator by the careful and meticulous observation of creation! That many today who so revere science and wrongly think it as adversarial to Truth and our Creator; don't even realize the science they believe in CAME FROM THE KNOWLEDGE of persons who KNEW AND OPENLY ACKNOWLEDGED OUR CREATOR! That SCIENCE itself COMES FROM OUR CREATOR; as well as the great minds who have recognized that fact in all of history.

The many fields of science ultimately owe their origins and continued existence to our Eternal, Intelligent Creator and to persons who; even though persecuted in varying degrees and some even murdered in the cause; still chose to Glorify GOD our Creator by the use of reason, rational thought, and meticulous observation of creation with lifelong meditation on the same! All who would be wise and knowledgeable do the same.

http://coldcasechristianity.com/2013/the-rich-historic-roll-call-of-great-christian-thinkers-and-scientists/

You have to understand that science today is based on FALLIBLE NOTIONS from FALLIBLE PERSONS, their observations may seem true in the moment, but more often than not they FAIL OVER TIME! **Therefore, PLACE YOUR TRUST IN GOD AND HIS WORDS IN THE HOLY BIBLE! God is not fallible; His Word is BOND! Absolutely TRUSTWORTHY!**
https://en.wikipedia.org/wiki/Superseded_theories_in_science_and_results_in_mathematics

This list is by no means exhaustive and is only the tip of the iceberg of all those who deserve to be recognized for their great achievements. If you are a scientist and by studying creation have discovered our Creator, I encourage you to write, publish, speak out about it; and try and monitor textbooks for our now faulty public indoctrination system; because we can save lives and souls from great suffering, great anguish; if we will only teach the children (and everyone) the Truth!

SO NEVER LET ANY IGNORANT, WICKED FOOLS, REMOVE THE MOST TRIED AND TESTED LITERATURE, THE HOLY BIBLE, CONTAINING THE MOST IMPORTANT FACTS OF HISTORY AND REALITY FROM OUR PUBLIC SCHOOLS, NEVER! PUT THE GREATEST REFERENCE BOOK FOR ACCURATE INFORMATION BACK IN THEM IMMEDIATELY! AND USE THE WORDS OF GOD TO TEACH CHILDREN OF ALL AGES EACH AND EVERY YEAR OF THEIR LIVES! (there is so much Divine Knowledge in the Holy Bible, that even if you study it thoroughly every day of your life on earth, you will NEVER learn EVERYTHING God is telling us therein! IT COMES FROM OUR ETERNAL CREATOR, HIS KNOWLEDGE IN JUST THAT ONE BOOK STILL SUPERCEDES THE COMBINED KNOWLEDGE OF MANKIND WORLDWIDE!) In other words, the Holy Bible has told us the future for millenniums before the future even came to pass, proving it has! It mentions technologies and events that they didn't even have words for in their ancient language, so the prophets described dreams and visions about these modern technologies in their own words, things like smart missiles, modern autos, jets, and more were all shown people who knew God THOUSANDS OF YEARS AGO!

SO YES, THE HOLY BIBLE IS A BOOK OF SCIENCE, TELLING US ABOUT SCIENTIFIC DISCOVERIES THAT ARE STILL COMING TO PASS!

Mathematicians have made astonishing discoveries:

https://www.khouse.org/articles/1995/102/

https://www.youtube.com/watch?v=WKHdyhmolMw&list=PLvmzfCYTsV-dnZPPlp4fhKB0nwipmrMka - Information Theorists and Cryptologists with the advent of modern computers, assert that the contents of the Holy Bible contain Divine Signatures, Messages from God, not just in the plain language of the words therein, but in the specific orders and meanings of the words and letters.

Some go so far as to believe that since each letter of the original text has numeric equivalences that there are mathematical formulas therein:

https://www.youtube.com/results?search_query=pi+and+e+in+the+bible

I go into further details on these subjects in my book on the fact that the Holy Bible is Provably and Verifiably Divinely Inspired. (I'm just mentioning some of the reasons in this book for those who haven't read it yet. It's in that book that I go into some of the prophecies about future events and technologies that have only come to pass millenniums later in recent history; before the eyes of the world today.)

Okay, REASON NUMBER ONE that the theory of evolution is false and impossible is because God Almighty says so quite plainly in the contents of the Provably and Verifiably Divinely Inspired Holy Bible.

I have written a book, The Holy Bible is Provably and Verifiably Divinely Inspired, in case you are not aware of the fact.

Genesis 1

English Standard Version

The Creation of the World

1 In the beginning, God created the heavens and the
earth. **2** The earth was without form and void, and darkness was over the face of the deep. And the Spirit of God was hovering over the face of the waters.

3 And God said, "Let there be light," and there was
light. **4** And God saw that the light was good. And God

separated the light from the darkness. 5 God called the
light Day, and the darkness he called Night. And there was
evening and there was morning, the first day.

6 And God said, "Let there be an expanse[a] in the midst of
the waters, and let it separate the waters from the
waters." 7 And God made[b] the expanse and separated the
waters that were under the expanse from the waters that
were above the expanse. And it was so. 8 And God called
the expanse Heaven.[c] And there was evening and there
was morning, the second day.

9 And God said, "Let the waters under the heavens be
gathered together into one place, and let the dry land
appear." And it was so. 10 God called the dry land
Earth,[d] and the waters that were gathered together he
called Seas. And God saw that it was good.

11 And God said, "Let the earth sprout vegetation,
plants[e] yielding seed, and fruit trees bearing fruit in which
is their seed, each according to its kind, on the earth." And
it was so. 12 The earth brought forth vegetation, plants
yielding seed according to their own kinds, and trees
bearing fruit in which is their seed, each according to its
kind. And God saw that it was good. 13 And there was
evening and there was morning, the third day.

14 And God said, "Let there be lights in the expanse of the
heavens to separate the day from the night. And let them
be for signs and for seasons,[f] and for days and
years, 15 and let them be lights in the expanse of the

heavens to give light upon the earth." And it was so. [16] And
God made the two great lights—the greater light to rule
the day and the lesser light to rule the night—and the
stars. [17] And God set them in the expanse of the heavens
to give light on the earth, [18] to rule over the day and over
the night, and to separate the light from the darkness. And
God saw that it was good. [19] And there was evening and
there was morning, the fourth day.

[20] And God said, "Let the waters swarm with swarms of
living creatures, and let birds[g] fly above the earth across
the expanse of the heavens." [21] So God created the great
sea creatures and every living creature that moves, with
which the waters swarm, according to their kinds, and
every winged bird according to its kind. And God saw that
it was good. [22] And God blessed them, saying, "Be fruitful
and multiply and fill the waters in the seas, and let birds
multiply on the earth." [23] And there was evening and there
was morning, the fifth day.

[24] And God said, "Let the earth bring forth living creatures
according to their kinds—livestock and creeping things and
beasts of the earth according to their kinds." And it was
so. [25] And God made the beasts of the earth according to
their kinds and the livestock according to their kinds, and
everything that creeps on the ground according to its kind.
And God saw that it was good.

[26] Then God said, "Let us make man[h] in our image, after
our likeness. And let them have dominion over the fish of
the sea and over the birds of the heavens and over the

livestock and over all the earth and over every creeping
thing that creeps on the earth."

27 So God created man in his own image,
in the image of God he created him;
male and female he created them.

28 And God blessed them. And God said to them, "Be
fruitful and multiply and fill the earth and subdue it, and
have dominion over the fish of the sea and over the birds
of the heavens and over every living thing that moves on
the earth." 29 And God said, "Behold, I have given you
every plant yielding seed that is on the face of all the
earth, and every tree with seed in its fruit. You shall have
them for food. 30 And to every beast of the earth and to
every bird of the heavens and to everything that creeps on
the earth, everything that has the breath of life, I have
given every green plant for food." And it was so. 31 And
God saw everything that he had made, and behold, it was
very good. And there was evening and there was morning,
the sixth day.

What we need to hear, with regards to origin of life on earth, is to identify KINDS of creatures and that **MANKIND was ORIGINALLY MADE IN THE IMAGE OF GOD!** Anyone claiming otherwise is out to lunch, no ifs, ands or buts. God tells us that each seed of every living plant has its OWN KIND! And that goes for every living sea creature, winged fowl, and beasts of the earth. THEY

NEVER CROSS OVER FROM ONE KIND TO ANOTHER! NEVER! The seed in them prevents this from happening!

If they do not breed, they are not the same kind! All creatures of the same kind, can breed and produce offspring! They can breed tall ones, short ones, hairy ones, etc. so there are great variety WITHIN THE KINDS of creatures. But NEVER does a dog kinds bread something like a bear kind! NO! not in the distant past! THAT'S all imaginary fantasy! No, there wasn't some animal that was once a fore runner of dogs and cats! Everyone saying so is just rambling on about their suppositions about prehistory with nothing but imaginations to go on! NOTHING IN RECORDED HISTORY OFFERS ONE SHRED OF HOPE to the evolutionary paradigm and INSTEAD ONLY OFFERS TO VALIDATE AND SUBSTANTIATE THE WORD OF GOD GIVEN US ALL IN THE HOLY BIBLE!

Genome

Reason Number Two of why, darwin's theory of evolution, is false and impossible is due to the fact genetics, genomes, of living organisms shows that is the case. No matter how much someone might want to believe in the theory of evolution, it doesn't ever bridge the gap between simple celled creatures and trillions of celled creatures; not anywhere in history. The genomes are so diverse that there is no natural method for going from a single celled amoeba to something like a fir tree through any proposed method of evolutionary thought, not in any stretch of the imagination, the genomes settle it, each living organisms is DESIGNED with their genetic makeup for that LIVING ORGANISM, they did not evolve. All such notions are just fabrications of deceived minds! Never let lunatics overtake the TRIED and TESTED contents of history! **Put the Holy Bible back into our schools NOW!**

https://www.youtube.com/watch?v=jSWwDCNNtrE - When you read scientific papers on the subject of embryonic development of various creatures and study the process under microscopes - https://www.youtube.com/watch?v=s-Xpa5UZAZs you find that all living organisms contain very specific DNA to them and their kind, such that different kinds cannot and do not reproduce. (darwin's theory of evolution is FALSE

AND IMPOSSIBLE because each and every living organism contains specific genetic information that is totally unique to that kind of creature.)

In other words, just what the Holy Bible plainly states: https://biblehub.com/hebrew/2233.htm that all living things have seed (genetic information) within themselves designed by God Himself to bring forth after its own kind/likeness from generation to generation. And that's what we observe ubiquitously and what modern genetics has redundantly proven. https://biblehub.com/interlinear/genesis/1-11.htm and

24And God said, "Let the earth bring forth living creatures according to their kinds: livestock, land crawlers, and beasts of the earth according to their kinds." And it was so. **25**God made the beasts of the earth according to their kinds, the livestock according to their kinds, and everything that crawls upon the earth according to its kind. And God saw that it was good.... https://biblehub.com/genesis/1-24.htm

In the original Greek "seed" regarding mankind was called "spermati"; and yes, we do derive our modern languages from words in the past. Yes, the Holy Bible is telling us that within all living organisms exists genetic information specific to that kind of living creature, that is perpetually designed by God Himself to bring forth like kinds unto itself; so never in any imagined amount of time does a single cell designed by Him to reproduce through asexual mitosis morph/evolve into our multitrillion celled selves that reproduce sexually. Never happened and never will, because as stated God Himself, in the contents of the Holy Bible, says so! And everything that is actually science only overwhelmingly has ever confirmed His Words as the Truth that they are.

Just ask yourselves the question if the information containing in each kind of genome is not specific to that kind of creature, then how can we even identify specific genomes at all? And if DNA isn't specific, then why is it used in court rooms to convict people of crimes? Billions and billions of nucleic acid sequences arranged themselves to form the shape of our bodies and other living things specifically. Genes just ORDERED themselves via evolutionary processes IS LUDICROUS and impossible. Not one thing mankind has made is made without ENGINEERING and therefore FUNCTIONS as we designed! Thinking that we ourselves just randomly formed all on our own without ENGINEERING to function as we do is preposterous!

The truth is that each kind of creature contains specific genetic information that reproduces ONLY that kind of creature from generation to generation. If someone argues that within the DNA of each kind of creature there is not information to form that kind of creature through each successive generation, then what tells that living organism to form into another like unto itself. It's in the genetics, the seed, just as God plainly has told us for millenniums and is only recently confirmed through the advent of many scientific and technological advancements. (Advancements that came about because of God, the Holy Bible, and those who studied His Words and His Creation.) SCIENCE ONLY EXISTS BECAUSE GOD AND THE HOLY BIBLE EXISTS AND IS TELLING US ALL THE TRUTH COVER TO COVER!

I've read enough papers on the topic to tell you plainly that there is no "junk DNA", just like there is nothing in your body that you don't need and is some kind of unnecessary part from their imagined evolutionary processes. From the moment you are conceived, your DNA is programming and coordinating every cell in your body to take the form of a boy or girl that grows into a man or woman. DNA is so complex that it not only takes proteins and other molecules to make each and every cell, but determines which of those cells form bones, which of them form muscles, which of them forms your organs and so on of your physical body. That kind of massive information coordinating trillions of cells to form your physical body is absolute evidence of "Intelligent Design", engineering! Or it wouldn't take "scientists" millenniums of human history to even discover it, let alone genetically modify anything.

If someone wants to jump off the deep end into insanity and irrationality and argue against Intelligent Design, don't jump with them or you look just as ignorant, deceived and self-deluded! DNA is INFORMATION and such intense and detailed information that even with supercomputers mankind still doesn't understand it all! If someone argues that DNA doesn't determine and coordinate each and every cell to form the various kinds of creatures, then why are genomes specific to the various kinds of creatures? (AGAIN, THE THEORY OF EVOLUTION IS FALSE AND IMPOSSIBLE BY EVERYTHING WE OBSERVE AND KNOW TO BE TRUE ALREADY!)

There are variations with each kind of creature such that they can produce other kinds of creatures like unto themselves, BUT there are NO POSSIBLE WAY one kind of creature EVER MORPHS into another kind of creature, without INTENTIONAL Genetic Modification. This is where EVOLUTION FAILS! Bottom line, all living things were INTELLIGENTLY DESIGNED BY OUR ETERNAL CREATOR.

If someone tries to argue against all reason that the universe created itself from nothing and life evolved despite everything we know to be true: horses breed horses, cows breed cows, apes breed apes, dogs breed dogs and so on and it's been that way throughout recorded history... i.e. no one plants seeds of corn and expects bananas or anything else but corn to grow as a result, then ask them to show their so called morphing organisms, ask them to show you in the fossil record anything evolving into anything else. It doesn't exist, never has and never will. ALL such claims are nothing but LIES. It would be similar today as if you took the bones of a dead pony and the bones of quarter horse and the bones of a massive shire and argued that either evolution or devolution has been proven just by arranging those bones in one sequence or another and claiming that they were all direct descendents either way. In other words, evolutionists are digging up fossils and telling their wild stories about prehistory, while they completely ignore the FACTS of history, well researched and thoroughly verified in the Holy Bible! They are ignoring that we do not see any beneficial mutations ANYWHERE in ALL HISTORY! and whatever birth defects of any organism that has ever been observed ARE DETRIMENTAL and causes such living creatures to die off. IN OTHER WORDS, what is called "junk DNA" is REDUNDANT PROGRAMMING that PREVENTS evolutionary claims, just as it is designed to prevent birth defects throughout the generations for all kinds of creatures. (The only mutations ever observed in all history resulted in harm to that creature, not anything beneficial.)

Yes, I have heard the brainwashed masses cry back at me, BUT YOU JUST DON'T KNOW THE THOUSANDS OF TESTS... blah, blah, blah. Yes, I've looked at all supposed claims of evolutionists that they imagine justifies their lunacy and cultic beliefs. NONE of those tests do so! Instead, they redundantly DISPROVE evolution. Like when you read an evolutionist going on and on about fruit flies or bacteria. And then coming to a totally FALSE conclusion based on the results. When fruit flies were subjected to all manner of environmental conditions, massive quantities of them died off, and any fruit flies that were physically changed were deformed, and unable to reproduce. When you poison a group of living organisms and a few of them survive, they haven't demonstrated evolution, they've only demonstrated that some living creatures can survive adverse environmental conditions. The man that survives war, can still die in a car accident. The bug that survives a pesticide can still be squashed, burned or eaten. Not one test has ever validated the notions of macro-evolution. Fruit flies after many thousands of generations never evolved into anything else, they were still fruit flies after deliberate efforts to force evolutionary madness upon them. Any and all experiments only show one kind of creature that always remains that kind of creature, it doesn't suddenly "evolve" into anything else.

Let's for arguments sake say that viruses (what God calls diseases) are factually mutating, even though in my own research, virtually all zoonotic pandemics in recent history, only occurred AFTER research began on those respective viruses. When you pull patents globally on all such research of these viruses from HIV to COVID, and everything in between, research began on those viruses BEFORE they ever spread globally and happened to coincide with deliberate genetic modifications of those viruses. Just because the wealthiest people on earth control governments and media is no reason to believe what you're being told by them. The same wealthy people have been caught on camera time and again, disdaining the poor masses, the people of the world as so far beneath them that Prince Philip was quoted saying:

https://www.google.com/search?client=opera&q=prince+philip+quote+of+coming+back+as+a+virus&sourceid=opera&ie=UTF-8&oe=UTF-8 - "In the event that I am reincarnated, I would like to return as a deadly virus, to contribute something to solving overpopulation."

So, when you look at their research and the facts, they spend trillions of dollars annually on the "military-industrial complex" sowing death and destruction to humanity worldwide, then it's more reasonable, in my opinion, to conclude the research in viruses, coupled with genetic modification of those respective viruses, followed by pandemics, is FAR MORE LIKELY to be deliberate bio-weapons, than anything "natural". It appears to me that at least one of the reasons they are pushing evolutionary nonsense so hard in the media and schools, is so that they can get away with these kinds of crimes against humanity.

That said, I know the Living God, and He told me that viruses such as AIDs would mutate because mankind wasn't receiving His Message, that we all must REPENT, turn from our wicked ways! Those diseases are still that kind of disease even if they become resistant to our attempts to eliminate them. That some diseases are part of the Divine Consequences that teach us NOT TO BE WICKED AND IMMORAL! (Deuteronomy 28) And so people trying to evade those REAL reasons for why they have become ill, are trying to cure themselves of those consequences SO THEY CAN GO ON SINNING! Instead of repenting of their wicked ways, and being healed inside and out by the Living God, Jesus Christ.

I still remember sometime after September 1988 and the beginning of 1992 reading front page news in some paper published here in America of a man who had been miraculously healed from AIDs. The man in the article said that the real disease wasn't AIDs but was separation from God and that after He had repented and received our Lord and Savior Jesus Christ, He had been completely healed with medical records as proof. YES! It happened! it made headlines! I spent some time looking through archives of the papers I thought it was in years later and unfortunately, don't remember which newspaper printed the article. For all I know the article might have been so controversial, that those who didn't like it tampered with the archives to omit it. (Similar to how they are tampering with content on the Internet and newscasts all over this world.) Yes, these extremely wealthy, evil people are KNOWINGLY lying to the public and yes, they do actively try to cover up their crimes and yes, they are deliberately attacking the masses not just by demoralizing us through the media and schools, but through false flags and many other methods of "population management".

(read my book on Population Management)

So, the presence of redundant DNA in each cell for each kind of creature is not "junk DNA" at all but is the

Divine Programming that coordinates the cells to form a dog, or a cat, or a man, or an ape, or a tree, etc. etc. etc. I call that present redundant Divine Programming in DNA, "form limiters" it's the information in every living kind that tells all those cells to be that specific kind of creature. As stated, the presence of redundant DNA in every cell CLEARLY PREVENTS evolutionary claims! (The theory of evolution is genetically impossible by what we observe and what we know is true.)

Lately, scientists have been studying fibroblasts, and embryonic stem cells, and deliberately inserting DNA into modified cells to form certain kinds of other cells. Again, this is not random evolution at all but is deliberate genetic modification, and while people claim they have good intentions, so far, all manner of genetic modification has only harmed life on earth.

http://www.blastthetrumpet.org/PublicLetters/AAAUpdatedPublicAlertsMattersofLifeandDeath/Updates053016/Strategies%20of%20War.pdf

http://www.blastthetrumpet.org/PublicLetters/AAAUpdatedPublicAlertsMattersofLifeandDeath/Updates053016/Crimes%20Against%20Humanity%20Part%201.pdf - part 1 http://www.blastthetrumpet.org/PublicLetters/AAAUpdatedPublicAlertsMattersofLifeandDeath/Updates053016/Crimes%20Against%20Humanity%20Part%202.pdf - part 2 of overt crimes against humanity

http://www.blastthetrumpet.org/PublicLetters/AAAUpdatedPublicAlertsMattersofLifeandDeath/Updates053016/Crimes%20Against%20Humanity%20Part%203.pdf - look people, YOU MUST KNOW GOD, and put GODLY PEOPLE IN POWER, because God has the answers for whatever you are facing in each generation until He Returns in Glory!

Population Growth Rates

The third reason why the theory of evolution is false and is making impossible claims, is population growth rates of all living organisms. In the Holy Bible, God tells us Eve is the mother of all living, and genetics has confirmed that all people living today come from one woman, dubbed "Mitochondrial Eve" but rather than see this as genetic confirmation for the history recorded in the Holy Bible, instead because of the evil God-denying money behind firing creation scientists and only employing ignorant, brainwashed people who are cultic members of evil-u-shun, they wrongly tell us that in the distant past many tens of thousands of years ago, their missing links all died off (and have never been found - any and all claims of apes to men evolution have been proven to be false, hoaxes or just improperly classified). I want people to understand worldwide that evolutionists are lying to themselves and you if they actually believe in it, because population growth rates of mankind and all creatures, defy their imagined beliefs that modern man has been on earth for more than the thousands of years of human history clearly recorded in the most thoroughly studied and verified book in all the world, the Holy Bible.

According to those who are funding abortion and other methods of death and depopulation after only a couple of millenniums mankind is "overpopulating" the earth and threatening to cause runaway extinction events, including our own. So no, modern men, "homo sapiens" have NOT been around longer than history plainly tells us in the Holy Bible and yes, anyone saying otherwise is defying the facts of observable reality such as population growth rates and the fact that the oldest living trees date commensurately to the Noahic deluge.

OBSERVED POPULATION GROWTH RATES OF ALL MANNER OF LIVING CREATURES WORLDWIDE TODAY, WOULD HAVE OVERPOPULATED THE PLANET SO MANY TIMES OVER THAT WE WOULD SEE MOUNTAINS OF THEIR REMAINS DIGGING ANYWHERE ON EARTH. (And no, we do not see, billions upon billions, or trillions upon trillions of human remains. Archaeological and anthropological evidence that has been found to date, overwhelmingly supports Biblical historicity and veracity.)

No one espousing MILLIONS to BILLIONS of years CAN EXPLAIN HOW AND WHY SUCH ORGANISMS CAN SURVIVE THAT LONG. Even conservatively, claims of hundreds of millions of years would have they're imagined DEEP TIME, take T-Rex for example, OVERPOPULATING THE WORLD MANY TIMES OVER! There is simply no way for mankind to have been here just for a few thousand years and already overpopulation is threatening extinction events to hear of their proposals; like a T-Rex, top predator, being around MILLIONS AND MILLIONS OF YEARS! It's all a complete fabrication on their part.

Imagine even the slowest amount of time for T-Rex and other top predators needed for population growth. Even their slowest time would have populated the world many MILLIONS of TIMES over during their deep-time evolutionary claims of hundreds of millions of years! Evolution is preposterously impossible!

There so-called geologic time of evolutionary claims, is just contrived because they can't think of how evolution could happen any other way! No way could their imagined epochs been around 100 million or more years, not even a million! CALCULATE THE POPULATION GROWTH RATES OF CREATURES TO SEE WHAT I MEAN, of not just the top predators, but of everything; it's all totally impossible nonsense of the worst kind!

(Read my book on the fact the Holy Bible is Provably and Verifiably Divinely Inspired - the Biblical Account is correct! That's the way it is folks!)

The Law of Biogenesis

The fourth reason the theory of evolution is false and impossible is the Law of Biogenesis. Yes, I've heard evolutionists screaming that their God and Bible denying theory has nothing whatsoever to do with the Origin of Life, but the Law of Biogenesis is not just refuting imaginary abiogenesis, but is refuting the claims of evolution.

https://bio.libretexts.org/Bookshelves/Microbiology/Microbiology_(Boundless)/01%3A_Introduction_to_Microbiology/1.01%3A_Introduction_to_Microbiology/1.1C%3A_Pasteur_and_Spontaneous_Generation - spontaneous generation, abiogenesis is not scientific and has been disproven.

The Law of Biogenesis plainly tells us all living organisms come from living organisms, all life comes from life! More specifically, any and all observations of living organisms throughout all history only produce like kinds, they never "evolve" or morph into anything but the kind of creature/Divine creation whose seed (genetic makeup) is in itself that it came from. (Again, validating the existence of our Eternal Creator, Jesus Christ, who plainly says all life comes from Him in the Holy Bible.)

Even if the eventuality occurs that life emerges from the seas and the lands it would STILL BE PROVING GENESIS! that GOD IS THE AUTHOR of all living things. It would in no way be supporting evolutionary thought! There are simply no failures of evolutions random mutations in the fossil record, so all life is intelligently designed! And despite they're drawings there are none of the intermarriages BETWEEN the kinds of creatures that allows for evolution, one KIND of creature MORPHING into ANOTHER KIND! It's just wrong and foolish notions by people who don't want to admit GOD DESIGNED US AND EVERYTHING!

Why does bubblehead thinking rule the day? It's because all those people DO NOT KNOW OUR ETERNAL CREATOR! Everyone who does not know our Eternal Creator really has no business in education until they do! Mankind is trying to figure out where life came from when our GOD came and explained it to them. He is the LIFE GIVER and He tells us He made all this specific to THEIR KIND. It would be one thing if a hippo mated with a giraffe and produced a hippo-giraffe but they don't. EVOLUTION IS IMPOSSIBLE NONSENSE! Everywhere God tells us HE PLACED SEED in LIKE KINDS of Creatures, those creatures can only produce LIKE KINDS OF CREATURES.

I want to restate and emphasize the fact that DNA specific to every kind of living organism, information so small and yet so vast that it STILL IS NOT FULLY UNDERSTOOD, encoded on protein molecules so small it takes our most powerful microscopes to even see them, is absolutely irrefutable proof of Intelligent Design, anyone denying that fact has completely left the realm of rational, reasonable thought! THINK! it is INFORMATION, INTELLIGENT INFORMATION, that is found within the SEED (sperm and egg) of a man and woman, that tells the "zygote" to begin producing the TRILLIONS of cells that ARE ALL COORDINATED by that microscopic INFORMATION to produce a human hand as opposed to the fin of a fish, a human heart as opposed to one of a bird, a human mind as opposed to one of a mouse, in other words EVERY SINGLE CELL in our bodies derives information from HUMAN DNA, found in the SEED, the chromosomes of HUMANS. If someone argues that genetics is not coordinating every single cell in our bodies with overt intelligent programming that tells those cells what their specific form and function is, then the question would remain what is? EITHER WAY, GENETICS COMBINED WITH THE LAW OF BIOGENESIS IS ABSOLUTELY CONFIRMING THAT ALL LIFE COMES FROM THE EXTREMELY INTELLIGENT GOD WHO HAS DECLARED HIMSELF TOLD US ACCURATELY ABOUT HIMSELF AND HIS CREATION IN THE CONTENTS OF THE HOLY BIBLE!

I repeat, either so much massive information is encoded on hydrocarbon protein molecules so small that our most powerful microscopes can barely see them, information that tells cells to form all the various kinds of creatures, from the simplest life forms all the way to the most complex (mankind), OR if genetic information is not responsible for coordinating billions and trillions of cells to all work together to form all the necessary parts, bones, muscles, organs, brains, each with DNA specific information on every cell particular to that kind of creature, then it would mean some manner of outside agency is at work forming each and every living creature. EITHER WAY, genetics is irrefutable proof of Intelligent Design because it takes intelligence to even perceive, identify and begin to comprehend its function. Genetic modification is further proof of Intelligent Design (it takes supercomputers and scientists deliberately modifying Divine Designs, genomes; showing us that INTELLIGENCE is behind all those DNA codes specific to every living creature), and so the INTELLIGENT DESIGNER is more specifically proof of the Eternal Living God who plainly declares Himself in the contents of the Holy Bible; (not chaotic, random, undirected, so-called "natural" processes; proposed by cultic believers in evolution at all.)

I'm referring to the fact God-denying atheists, cultic believers in evil-u-shun scream for separation of church and state (even though that phrase means the

EXACT OPPOSITE of what they imagine it means, namely that the state is not to interfere with our rights to practice our religion and free speech anywhere in these United States. (Bill of Rights Number 1 - freedom OF religion (not FROM it) and freedom OF speech, the right to not only believe in God with a clear conscience but to speak about Him in any venue in these United States! Bill of Rights Number 2 - the right to bear arms (specifically follows Number 1 so that all citizens KNOW with certainty, that they should fight to the death to protect Number 1!) In other words, what God-denying wicked people are doing along with treasonous judges are showing themselves to be spitting upon our most important rights, fought, bled and died for by our founding fathers! AND I SAY THEY ARE DOING THIS MOST HYPOCRITICALLY, because as this and virtually everything I write about plainly shows, it's THEM! the God-denying fools that are engaging in blind, unproven, unscientific cultic beliefs! While Biblical Creation Scientists have ALWAYS based their worldview on the MOST IMPORTANT FACTS of history and observable reality! (so cultic believers in evil-u-shun have been censoring history and true science, by wrongly saying anything "religious" has no place in our public schools, (even though our public schools were founded by Holy Bible believers) and by attempting to twist "separation of church and state" to mean the exact opposite of what the highest Laws in our nation plainly state! and what the original mention of the phrase clearly indicates.)

I REPEAT! What we have learned about genetics is that genetic information SPECIFIC TO EACH KIND OF CREATURE IS PROGRAMMED INTO the seed, the reproductive method, of each and every kind of creature. In the case of MANKIND, genetic information SPECIFIC to MANKIND exists in the seed(sperm) of the man, and egg(ovum) of the woman, such that MANKIND ONLY PRODUCES MANKIND from generation to generation! When the SEED (sperm) of the man and SEED (ovum) of the woman unite, that genetic information COMBINES, pairs together BECAUSE that genetic information is a MATCH! ONLY LIKE KINDS OF CREATURES CAN REPRODUCE! (just as the Words of our Eternal Creator plainly state in the very first chapter of the provably and verifiably Divinely Inspired Holy Bible!) And when that paired match of specific genetic information combines, that "zygote" then passes on that SAME SPECIFIC GENETIC INFORMATION TO EVERY CELL THEREAFTER, SUCH THAT EVERY CELL IN YOUR BODY IS SPECIFICALLY A CELL OF MANKIND! (no other creature!) In other words, TRILLIONS OF TIMES OVER, every cell says I AM a CELL that belongs specifically to MANKIND! (no other creature) and that is true about all living organisms! Every cell in a horse, tells us genetically speaking that it belongs specifically to that kind of creature as opposed to any other kind of creature and so on! SO, ALL CREATION IS SPECIFICALLY PROGRAMMED BY OUR ETERNAL CREATOR, JESUS CHRIST, TO SPECIFICALLY PRODUCE ALL THE VARIOUS KINDS OF CREATURES FROM GENERATION TO GENERATION. You can take a genetically modified cell and reproduce it in another creature, but in no way happens by evil-u-shun. So, if any genome was altered say as to produce a mutation or deletion of Divine Information therein, IT IS ONLY ALWAYS HARMFUL, leading to the death or eventual extinction of such mutated creature, and such mutations and deletions of genetic information ARE NEVER IN ANY WAY SUFFICIENT to go from a single celled bacterium, or any other single-celled simple creature like an amoeba, into our multi-trillion celled selves NOT IN ANY AMOUNT OF IMAGINED TIME! **Evolution is plainly FALSE and IMPOSSIBLE, by what we know to be certainly true!**

Some argue that the flood of Noah was necessary because evil spirits, fallen angels, wicked gods, had corrupted genetic information, Divine Designs, on earth such that all flesh (genetic information therein) had been corrupted and abominations were resulting, like "giants" or "nephilim". That were thoroughly evil and so evil it warranted destroying virtually all life on earth to get rid of that genetic abomination. It's for this reason, I am against all manner of genetic modification of anything: simply put, mankind doesn't really understand the full implications of their actions and none of us know better than our Eternal Creator. Intentionally modifying His own Engineering of Life, has been proven to be only harmful. Mankind is like little kids playing with a biological bomb... I wonder what happens if we do this? I wonder what happens if we do that? until at the last it explodes in their faces because they were simply too ignorant to realize how dangerous it was. **Crimes against humanity and life on earth are taking place, not just deliberately, but out of a massive lack of knowledge. EVERYONE AND I MEAN EVERYONE, NEEDS TO KNOW AND LEARN FROM OUR ETERNAL CREATOR, LORD AND SAVIOR, JESUS, THE CHRIST!**

http://www.blastthetrumpet.org/PublicLetters/AAAUpdatedPublicAlertsMattersofLifeandDeath/Updates053016/Crimes%20Against%20Humanity%20Part%201.pdf - crimes against humanity are being committed because some people imagine they know better than God Almighty and ignorantly do things they don't realize are harmful to us and life on earth.

http://www.blastthetrumpet.org/PublicLetters/AAAUpdatedPublicAlertsMattersofLifeandDeath/Updates053016/Crimes%20Against%20Humanity%20Part%202.pdf - When people attempt to control the weather, or modify living organisms, they are actually doing things contrary to the best interests of life on earth, all while imagining otherwise.

http://www.blastthetrumpet.org/PublicLetters/AAAUpdatedPublicAlertsMattersofLifeandDeath/Updates053016/Crimes%20Against%20Humanity%20Part%203.pdf - we don't want evil people in power and we don't want greedy, apathetic people in power, we need the most honorable, intelligent, conscientious people in power (people who KNOW GOD and are learning straight from Him, by the Presence of His Holy Spirit!) Otherwise, there is a lack of proper actions taking place on earth and evil ones instead due to ignorance and wickedness.

So, when antichrists, like Rockefeller, ripped public education away from the sane and educated populace (Holy Bible believing Christians) and began brainwashing innocent children with lies, it has only resulted in a nation of increasingly ignorant, deceived and self-deluded persons and should it continue threatens to erode civilization, law and order and end in widespread crime and violence. It's the reason God tells us He gives us a sound mind in the scriptures, one that acknowledge Him, His Words in the Holy Bible as absolute truth and any imaginations contrary as wrong. (It has always been that way and always will be, because God Almighty is the Ultimate Authority in all Creation.)

Bottom Line: America and all nations on earth need to put the Holy Bible in all schools and universities as the primary textbook for accurate information in all the world; the very foundation of truthful knowledge that it is.

https://www.nps.gov/articles/000/church_state_historical.htm - the first mention of the phrase "separation of church and state" confirms what the Bill of Rights tells us all so plainly: namely the state is NOT TO INTERFERE with our rights of religion and speech! (so, it was never intended to kick the mention of God and the Holy Bible out of the schools that the People of God and the Holy Bible FOUNDED IN THE FIRST PLACE!) Bottom line, if you want to be a wicked, ignorant, deceived and self-deluded fool (God-denying person who has no rationale for reality before their eyes; origin of the universe and life), then don't go to schools that taught directly from the Holy Bible, because the One True God, plainly commanded in its contents to teach children from every generation about Him and His Commandments therein!

https://www.freedomforum.org/separation-of-church-and-state/ - this phrase was NEVER to censure any of us from speaking and teaching what is TRUE! (whether it is considered to be "religious" or not!) That's why I'm pointing out that unless you are God Almighty Himself, every one of us (including those throwing tantrums saying they are NOT religious) are by definition holding beliefs that others can easily consider to be such. (in other words, atheists, are some of the most religious persons on the planet! they SCREAM we are IRRELIGIOUS with religious fervor, and AGGRESSIVELY try to censure any religious speech or displays anywhere! they are demonstrating cultic beliefs without any rational support whatsoever. I say cultic, because they idolize ONLY GOD-DENYING fools like themselves! and imagine their own imaginations and that of their fellow GOD-DENYING FOOLS are superior to GOD ALMIGHTY, our Eternal Creator, when every single one of them are believing things that are simply not true or they would all be theists. **(The atheist has no rational for reality before our very eyes, plain and simple; no ifs, ands, or buts; and actively dismiss the most important facts of history that all tell us who the One True God is and how to know Him and learn from Him personally.)**

We need to be concerned with teaching children those FACTS, TRUTHFUL INFORMATION, instead of attempting to censor any mention of anything deemed to be religious in our schools. BECAUSE censoring "religious" information requires rewriting all of history, which means teaching children LIES! and it requires telling them stories (like the theory of evolution and how they descended from apes and pond scum) in the name of science, rather than telling them the well verified truth, that the Holy Bible has been and always will be the most accurate reference book for truthful information on the planet! The FOUNDATION of literary TRUTH laid by GOD ALMIGHTY HIMSELF and given to all mankind!

Yes, I have read nonsense from arrogant persons that aren't ashamed to display their ignorance; that falsely claim the Law of Biogenesis says nothing about abiogenesis, I call them arrogant and ignorant because both the Law of Biogenesis and abiogenesis cannot be true, it's either one or the other BY DEFINITION! Get it through your thick skulls! the Law of Biogenesis, life from life, living organisms from living organisms, soundly disproves any notions of abiogenesis, spontaneous generation!

Biogenesis - LIFE COMES FROM LIFE or a-biogenesis - NOT biogenesis! All living come from something living or it doesn't! both the observed and well verified scientific Law of Biogenesis and imaginary abiogenesis CANNOT BE TRUE. it's one or the other! **So yes, the Law of Biogenesis throws a massive wrench in the imaginary cogs of the God-denying, evolutionary paradigm! IMPOSSIBLE! NOT SCIENCE!**

whether the medical definitions or the general definitions are used:

Prefixes, medical: Medical words are often put together, cobbled from two or more building blocks. Among these building blocks are the prefixes.

Examples of prefixes used in medicine include:

- a- : Prefix much employed in the health sciences indicating "not, without, -less" as, for example, in alexia (not read), aphagia (not eat), aphonia (not voice, voiceless). The "a-" usually becomes "an-" before a vowel as, for example, in anemia (without blood), anotia (no ear), anoxia (no oxygen). The prefix "a-" comes from the Greek meaning "not."

https://membean.com/rootcasts/a-not/ - clearly, BY DEFINITION Biogenesis and a-biogenesis are antithetical; cannot both be true. i.e. **LIFE comes from LIFE supports what God tells us in the Holy Bible; not the evolutionary paradigm wrongly being taught in the name of science currently**. He tells us that He formed all things and gave them seed within themselves to BRING FORTH LIKE KINDS OF CREATURES so NONE OF THEM MORPH over time INTO ANY OTHER KIND OF CREATURE! Mankind is in the image of GOD! not monkeys and pond scum!

John 5:39-40

39 You search the Scriptures because you think that in
them you have eternal life; and it is they that bear witness
about me, 40 yet you refuse to come to me that you may
have life.

John 14:6-9

6 Jesus said to him, "I am the way, the truth, and the life. No one comes to the Father except through Me.

The Father Revealed

7 "If you had known Me, you would have known My Father also; and from now on you know Him and have seen Him."

8 Philip said to Him, "Lord, show us the Father, and it is sufficient for us."

9 Jesus said to him, "Have I been with you so long, and yet you have not known Me, Philip? He who has seen Me has seen the Father; so how can you say, 'Show us the Father'?

The One who has Eternal Life tells us in Genesis that He is the One who Created and Made all manner of living creatures. Jesus Christ said He is our Eternal Creator and proved it beyond reasonable doubt. He also told us all how to know and learn from Him personally; so, no one has any excuse.

(again, read my book, The Holy Bible is Provably and Verifiably Divinely Inspired) https://www.youtube.com/watch?v=QmiwQ0Kosa8 - here's a fine example of a cultic believer in evolution looking at evidence that his beliefs are NOT true, all while claiming they are. In other words, all of the genetic tests he cites that have been done throughout history have proven pea plants produce pea plants, foxes breed foxes, fruit flies produce fruit flies, mice breed other mice and so on so forth; no matter how many generations of those living creatures pass. So, within the various kinds of creatures, there is enough Divine Programming to get big ones, small ones, tall ones, short ones, fat ones, skinny ones and so on; VARIATIONS WITHIN THE LIMITATIONS OF THEIR DESIGNED DNA, but NO WAY TO BREAK THROUGH to the madness of claiming ancient bacteria or any single celled organism "evolved" into complex plants and animals, or anything else but bacteria over time. VARIATION WITHIN THE KINDS IS NO JUSTIFICATION WHATSOEVER TO CONCLUDE "goo to you, by way of the zoo" madness; that ancient single cells "evolved" both into bananas and the monkeys that eat them over "billions" of imagined years. **The theory of evolution is false, impossible, ludicrous and most unscientific! Everything we know about the facts of history and observable reality soundly disproves it.**

Science, the facts of history, the accumulated knowledge of mankind, DOES NOT support the claims of evolution or virtually any of the gibberish being proposed

today from those who are denying our Eternal Creator. Almost all of what they're saying about the origin of the universe, and life on earth, needs to be properly discarded, removed from the media and our public schools and the Holy Bible put back in immediately. Believing in imaginary nonsense with regards to important subjects as to who we are, where we came from and where we're going, is shameful and harmful; you are NOT an "evolved" animal; you are NOT descended from apes or monkeys or ancient "ape-like" creature! **You are a Divine Creation with Divine Purpose, created and made in the Image of God Almighty, Jesus Christ, Himself! God is real, everyone, everywhere NEEDS to KNOW Him for their own good and the good of all creation.**

When you place the stories of the God-deniers next to the Biblical Creation Account the Biblical Account is THOROUGHLY SUPPORTED AND TRUE BY EVERYTHING WE KNOW TO DATE! while the story being wrongly taught in the name of science currently isn't! **Attempting to remove truthful information from our schools just because some deem that information to be "religious", has only caused facts to be replaced with fiction, very sad nonsense, instead.** So cultic lies are what is being taught today in our schools, turning children sadly into a bunch of people who have been thoroughly brainwashed and so much so that they are resistant to truthful information. Even when I show them facts, they either refuse to look at those facts, or comprehend them. It's exactly as **Yuri Bezmenov** stated about demoralization/brainwashing does to people:

https://youtu.be/Hr5sTGxMUdo?t=312 - once a person has been successfully demoralized/brainwashed they are virtually impossible to persuade with truth and facts WHICH IS WHY IT'S IMPERATIVE THAT UNTIL NATIONS WISELY PUT THE HOLY BIBLE IN SCHOOLS, godly parents MUST TEACH THEIR CHILDREN AT HOME! DO NOT ALLOW GOD-DENYING FOOLS to LIE to your children! to brainwash them with nonsense! It makes them look like idiots, instead of properly educated intelligent adults! People ARGUING they are a direct descendant of some ape-like creature in the distant past! IMAGINE! how many monkey-men would it take to create one of them? THINK! It would be LAUGHABLE if not so sad, anyone has been so thoroughly brainwashed they actually believe that incredible nonsense! Not only has not one "missing link" ever been found, but any and all claims of them have been found to be either mistakes or deliberate hoaxes:

https://www.youtube.com/watch?v=-ElpINS3zU0 - lies are being taught innocent children, getting them to believe in things that are most definitely NOT TRUE! (like the disproven, false and impossible "theory of evolution") I encounter brainwashed people that laugh at Kent Hovind, even though Hovind is telling them the plain truth! It's SAD! brainwashed people can't tell fact from fiction! truth from lies! So, we MUST STOP sending children to public schools UNTIL our schools put TRUTH and the Holy Bible BACK IN THEM! Creation Science is being wrongly censored by frankly, either deliberately evil antichrists, or brainwashed, deceived persons! ALL SCIENCE ORIGINATED FROM BIBLICAL CREATIONISTS! IT NEEDS TO REMAIN THERE! ANYONE DEPARTING FROM THE CONFINES OF THE TRUTH ALREADY LAID BY GOD ALMIGHTY AND HIS WORDS IN THE HOLY BIBLE ARE THE ONES THAT SHOULD BE FIRED! NOT BIBLICAL CREATIONISTS!

https://www.youtube.com/watch?v=_3CQtIWnx7c - because we have cultic believers, brainwashed persons or very evil ones who know they are lying to children, currently in power, they are censoring TRUTHFUL information that soundly and redundantly disproves the nonsense of the theory of evolution!

SO, LIES ARE DELIBERATELY BEING TAUGHT LITTLE KIDS IN OUR PUBLIC SCHOOLS ALL THE WAY UP TO YOUNG ADULTS IN OUR UNIVERSITIES! ALL BECAUSE OF GOD-DENYING, GOD-HATING WICKED PEOPLE! WHO WOULD RATHER LIE THAN FACE THE TRUTH! Namely God is most definitely real, and He is the God of the Holy Bible, Jesus Christ!

The Fossils

This leads me to the fifth reason of how and why we know the theory of evolution is false; fossils buried in worldwide sedimentary layers, with marine fossils even on the top of Mount Everest. All fossils (especially when over 90% of them are all marine) in global sedimentary layers are best explained by the historic flood; not any of the deep time claims currently being proposed by those who are in denial of the facts of history, like the Noahic deluge. **(Again, if you don't understand that not one God-denying person has the credibility of Jesus Christ, the One True God; you need to read my book, The Holy Bible is Provably and Verifiably Divinely Inspired.)**

It's mind boggling to me how I can point to reality before our eyes and tell the truth that it is all evidence for our Eternal Creator, but the atheist will point to the same evidence and say it's proof of their imagined tale of evolution. In other words, atheists will say in one breath that reality is evidence... just not evidence of God or Biblical historicity and veracity, even though it is. Again, examine the two stories side by side! detail by detail!

https://www.youtube.com/watch?v=UM82qxxskZE - **the Biblical Creation Account is thoroughly supported by**

everything we see and know about reality! THOROUGHLY!

evolution isn't! - https://www.youtube.com/watch?v=GjvuwneORrE

https://firstamendment.mtsu.edu/article/establishment-clause-separation-of-church-and-state/ - I want to repeat that in these United States and throughout western civilization, public education came into existence because of the God of the Holy Bible; through the Church of Jesus Christ, people who believe in God and His Words in the Holy Bible. Therefore, SINCE THAT STATEMENT is a PROVEN FACT OF HISTORY, the State had/has NO RIGHT WHATSOEVER to take public education away from the Bible believers who founded it! **NONE! Public education in American and all over this world, needs to continue to be in the hands of Bible believers! NO ONE ELSE!** If you want to teach your children proven lies and fictional beliefs, that is an evil you must bear all on your own privately, but you cannot force your lies to be taught children created and made by the One True God, the God of the Holy Bible, Jesus Christ! **AMERICANS AND ALL PEOPLE NEED TO BE MORE CONCERNED ABOUT TEACHING CHILDREN TRUTH, THAN THEY ARE ABOUT CENSORING TRUTH, JUST BECAUSE TRUTH FALLS INTO**

THE RELIGION OF CHRISTIANITY, THE GOD OF THE HOLY BIBLE! TRUTH IS TRUTH! and plainly the Truth is GOD, Jesus Christ! (John 14:6-9; 8:32-36 and for those who whine on with, "that's only a claim." John 14:20-26 - it's verifiable! so NO ONE has any excuse!)

To be sure, the founders of America, wanted the freedom for everyone to live according to their God-given conscience, and didn't want to sponsor any specific denomination of Christianity, but they most definitely did NOT intend, for madness to ensue, such as imagining all religions are equal, lawful or acceptable, or that only the irreligious should be teachers or politicians. Our founding documents mention God throughout and that it's impossible to maintain government, or law and order without God and a godly citizenry! (and YES! it is referring to the God of the Holy Bible! as the Supreme Court properly examined and identified! - https://firstamendment.mtsu.edu/article/church-of-the-holy-trinity-v-united-states/#:~:text=Congress%2C%20public%20domain)-,Church%20of%20the%20Holy%20Trinity%20v.,is%20a%20“Christian%20nation .”) So, our laws were not intended to in any way support departing from the One True God, the God of the Holy Bible! not all religions are lawful, and therefore cannot be lawfully practiced!

(read my book on "Save the World from islam", including muslims) - so NO! our laws are not for ungodly, unlawful, religious or irreligious people, such persons are criminals, and arguably insane! and NO! our laws do not protect criminal organizations masquerading as a religion; like islam! They were strictly intended to a continual mention of the One True God, the God of the Holy Bible, and not to establish any denomination thereof! Never did Holy Bible believers that fought for and established these United States imagine people would depart so far from the facts of history and truth, that they would be twisting their own words and laws to be used against them! (such as organizations like freedom from religion, which is most definitely a subversive and treasonous organization, twisting our foundational laws of freedom OF religion (NUMBER ONE ON THE BILL OF RIGHTS) to mean just the opposite.)

Our founders reasoned that as long as Christians held the right to religion and the right of free speech, to speak according to our God-given consciences, that the Truth would always prevail! BUT NOT WHEN irreligious traitors in our midst have been ripping those rights away from us! Firing Biblical Creationists from our schools and universities, replacing it with their imaginary godless dogmas; brainwashing and deceiving innocent children away from the Truth! LYING TO THEM DELIBERATELY and it appears with all the many sexual scandals, that those many lies and the deliberate act of censoring the Holy Bible from public education is specifically for the purpose of corrupting those children due to their personal depravity and sexual immorality! In other words, traitors and enemies of America as it was founded to be are attacking America and Americans within our own borders! And as a result, we now have the kind of people God drowned and burned in the past, actually proud of their immorality and depravity, threatening to lead our own nation into such ruin!

http://www.blastthetrumpet.org/PublicLetters/AAAUpdatedPublicAlertsMattersofLifeandDeath/Updates053016/America%20is%20Under%20Attack.pdf - **anyone in our borders departing from God, the One True God, who plainly tells us who He is and how to know Him in the contents of the Holy Bible, is an enemy of America and if native born, a subversive traitor!**

https://www.johnadamsacademy.org/apps/pages/index.jsp?uREC_ID=2003858&type=d&pREC_ID=2094472 - "**#2 – Public and Private Virtue**

John Adams said, "Our constitution was made only for a moral and religious people. It is wholly inadequate to the government of any other."[1] Morality and virtue are the foundation of our republic and necessary for a society to be free. Virtue is an inner commitment and voluntary outward obedience to principles of truth and moral law. Private virtue is the character to govern oneself according to moral law at all times. Public virtue is the character to voluntarily sacrifice or subjugate personal wants for the greater good of other individuals or the community. Specific moral virtues include charity, justice, courage, temperance, reverence, prudence, and honesty. These virtues are the moral fiber and moving force to act in accordance with wisdom. Our scholars embrace these virtues and seek to incorporate them in the John Adams Academy community through which these virtues are cultivated and practiced.

[1] John Adams, "Letter from John Adams to Massachusetts Militia," 11 October 1798."

BOTTOM LINE: departing from God, Jesus Christ, and His Commandments given us all in the Holy Bible tears away at the very foundation of law and order and threatens to lead all those doing so into widespread anarchy, suffering and death!

https://oll.libertyfund.org/pages/1776-witherspoon-dominion-of-providence-over-the-passions-of-men-sermon - "I shall now conclude this discourse by some exhortations to duty, founded upon the truths which have been illustrated above, and suited to the interesting state of this country at the present time; and,

1. Suffer me to recommend to you an attention to the public interest of religion, or in other words, zeal for the glory of God and the good of others. I have already endeavored to exhort sinners to repentance; what I have here in view is to point out to you the concern which every good man ought to take in the national character and manners, and the means which he ought to use for promoting public virtue, and bearing down impiety and vice. This is a matter of the utmost moment, and which ought to be well understood, both in its nature and principles. Nothing is more certain than that a general profligacy and corruption of manners make a people ripe for destruction. A good form of government may hold the rotten materials together for some time, but beyond a certain pitch, even the best constitution will be ineffectual, and slavery must ensue. On the other hand, when the manners of a nation are pure, when true religion and internal principles maintain their vigour, the attempts of the most powerful enemies to oppress them are commonly baffled and disappointed. This will be found equally certain, whether we consider the great principles of God's moral government, or the operation and influence of natural causes.

What follows from this? That he is the best friend to American liberty, who is most sincere and active in promoting true and undefiled religion, and who sets

himself with the greatest firmness to bear down profanity and immorality of every kind. Whoever is an avowed enemy to God, I scruple not to call him an enemy to his country. Do not suppose, my brethren, that I mean to recommend a furious and angry zeal for the circumstantials of religion, or the contentions of one sect with another about their peculiar distinctions. I do not wish you to oppose any body's religion, but every body's wickedness. Perhaps there are few surer marks of the reality of religion, than when a man feels himself more joined in spirit to a true holy person of a different denomination, than to an irregular liver of his own. It is therefore your duty in this important and critical season to exert yourselves, every one in his proper sphere, to stem the tide of prevailing vice, to promote the knowledge of God, the reverence of his name and worship, and obedience to his laws."

I only mention these quotes to let you know that teaching the lies of evolutionary nonsense in our schools to innocent children threatens the viability of our nation! We cannot depart from God and His Precepts in the Holy Bible without dire consequences! It's not just fundamentally wrong to teach children lies in the name of science, but it's un-American, unpatriotic, unlawful and downright evil; so evil that it threatens to result in not just civil unrest but the dissolution of our once great nation, if it is allowed to continue! **THE MANY LIES OF**

EVOLUTIONARY NONSENSE MUST BE REMOVED FROM ALL PUBLIC INSTITUTIONS OF LEARNING, AND THE TRUTH, THE MOST IMPORTANT FACTS OF HISTORY CONTAINED IN THE HOLY BIBLE PUT BACK IN IMMEDIATELY!

BOTTOM LINE: Public education exists because of God, His Words in the Holy Bible, and His People; secularists have no right whatsoever to take it away in order to brainwash kids with their many lies! NONE! And frankly, is an act of open treason taking place within our borders by any and all wicked, God-denying persons doing so.

All persons believing lies, like the theory of evolution, need to let this truth sink into their presently deceived minds, the Holy Bible has been right about all things within its contents when vain opinions by so-called scientists have been proven wrong again and again and again all throughout history! (the theory of evolution is no exception) SO NEVER, EVER believe the opinion of ANYONE over the tried and tested, thoroughly researched and well verified facts, in the Holy Bible! NEVER! God, Jesus Christ, doesn't lie! His Words are Truth! And the standard by which ALL other opinions from ANYONE are tested by!

Christ Our Foundation

10By the grace God has given me, I laid a foundation as an expert builder, and someone else is building on it. But each one must be careful how he builds. **11**For no one can lay a foundation other than the one already laid, which is Jesus Christ. **12If anyone builds on this foundation using gold, silver, precious stones, wood, hay, or straw, 13 his workmanship will be evident, because the Day will bring it to light. It will be revealed with fire, and the fire will prove the quality of each man's work. 14If what he has built survives, he will receive a reward. 15If it is burned up, he will suffer loss. He himself will be saved, but only as if through the flames. https://biblehub.com/1_corinthians/3-15.htm**

Bottom Line: all lies will come to an end, only Truth remains.

33 Heaven and earth will pass away, but My words will never pass away. https://biblehub.com/luke/21-33.htm

So NEVER trust the vain opinions of ANYONE (scientist or otherwise) above the Words of God in the Holy Bible! NEVER!

Again, read my book "The Holy Bible is True: Provably and Verifiably Divinely Inspired"

https://youtu.be/qERdL8uHSgI - it's presentations like this that I call misinterpretation of the data before us all. Throughout the world today there are still anaerobic organisms:

https://www.google.com/search?client=opera&q=ANAEROBIC+BACTERIA&sourceid=opera&ie=UTF-8&oe=UTF-8 - so no, anaerobic creatures have not gone extinct billions of years ago by the presence of aerobic ones. All manner of life forms are all over this planet, even some that live without sunlight at the bottom of our oceans:

https://en.wikipedia.org/wiki/Hydrothermal_vent#Biology_of_hydrothermal_vents

So, what I mean by misinterpreting the data of the Fossils found in the sedimentary layers is that even living organisms are already found in various locations and environments still today. And so, when marine fossils are found even on the top of Mt. Everest and throughout all those layers down to the very bottom of them, that is evidence of the flood! evidence of liquefaction, hydrologic sorting, and simply burying all manner of living organisms at once. We find the smallest fossils of unicellular organisms THROUGHOUT the sediments, not EXCLUSIVELY at the bottom but if I take marine mud and pour it over "dinosaurs" then at the bottom that mud would contain single cells, followed by slightly larger marine organisms and so on and yes herds or concentrations of certain life forms would be buried together, just as the evidence shows. Finding marine invertebrates buried with reptiles and mammals altogether is no justification whatsoever for the imaginary geologic column or ages and instead is absolutely evidence of the flood. **GOD HIMSELF SAID THE FLOOD OCCURRED SO IT'S A FACT and the evidence shows it!**

Over and over again the deep time claims of evolution have been shown to be FALSE! Whenever we read history of "dragons" or see them depicted on ancient stones or architecture, people today call that history, "myths and legends", when none of that history is written as such and yet whenever they talk about BILLIONS of years ago... "it's scientific fact!" GET YOUR HEAD ON STRAIGHT FOLKS, history is history, all those notions about billions of years of prehistory aren't!

In other words, there are plenty of creatures still alive today that would be called dinosaurs or dragons: creatures like alligators, crocodiles, various lizards, including... wait for it... komodo dragons. So, we can see the mention of dragons in the past were not all works of fiction at all, just history.

https://www.google.com/search?client=opera&q=evidence+that+humans+and+dinosaurs+coexisted&sourceid=opera&ie=UTF-8&oe=UTF-8 - do some research and you will see that dragons/dinosaurs are mentioned as facts of history in some ancient texts, as well as engraved on stones details like texture of skin, and found engraved on some ancient temples from various ancient cultures. All I'm saying is those who are adamant about the prehistoric tales being taught in the evolutionary paradigm are not backed by the recorded facts of history and observable reality; the history in the Holy Bible is.

Here's a five-part series I wrote on why the entire evolutionary paradigm being taught in the name of science today just isn't science; it's not even decent science fiction:

http://www.blastthetrumpet.org/PublicLetters/AAAUpdatedPublicAlertsMattersofLifeandDeath/Updates053016/Th e%20Modern%20Evolutionary%20Paradigm%20is%20NOT%20SCIENCE%20part%201.pdf - part 1 of why the entire evolutionary paradigm currently wrongly being taught in the name of science isn't scientific at all... it doesn't even rise to the merit of decent science fiction!

http://www.blastthetrumpet.org/PublicLetters/AAAUpdatedPublicAlertsMattersofLifeandDeath/Updates053016/The%20Modern%20Evolutionary%20Paradigm%20is%20NOT%20SCIENCE%20part%202.pdf - part 2 virtually all the theories of science regarding origin of our universe and macro-evolution ARE IMPOSSIBLE unfortunately people just IMAGINE otherwise. WE ALL HAVE TO LEARN FROM GOD! He tells us the TRUTH in the contents of the Holy Bible!

http://www.blastthetrumpet.org/PublicLetters/AAAUpdatedPublicAlertsMattersofLifeandDeath/Updates053016/The%20Modern%20Evolutionary%20Paradigm%20is%20NOT%20SCIENCE%20part%203.pdf - part 3 - **There are plenty of authors who have debunked the theory of evolution, and support Intelligent Design, PEOPLE SHOULD BE LISTENING TO THEM!**

http://www.blastthetrumpet.org/PublicLetters/AAAUpdatedPublicAlertsMattersofLifeandDeath/Updates053016/The%20Modern%20Evolutionary%20Paradigm%20is%20NOT%20SCIENCE%20part%204.pdf - part 4 - **the deep time claims have been redundantly disproven by all the soft tissue dinosaur specimens. their entire fabricated**

background for support the evolutionary paradigm over millions and even billions of years is NON-EXISTENT.

http://www.blastthetrumpet.org/PublicLetters/AAAUpdatedPublicAlertsMattersofLifeandDeath/Updates053016/The%20Modern%20Evolutionary%20Paradigm%20is%20NOT%20SCIENCE%20part%205.pdf - part 5 - like it or not HISTORY is telling us the TRUTH, yes, the HISTORY recorded in the Holy Bible.

This is what we are teaching now in our schools; it is hogwash of the worst kind! None of it is true!

The Modern Evolutionary Paradigm is NOT SCIENCE, it isn't even good Science Fiction part 1 of 5 November 22, 2014 at 10:06am Public

In the Holy and Omnipotent Name of YAHOSHUAH (JESUS CHRIST), I command the spirit of error and all foul and unclean spirits to depart from all who read these words. I call for the spirit of Repentance to come upon all who read these words and May GOD our FATHER grant His Holy Spirit to all who read these words and fulfill what is written in Acts 2:38 – 38Peter said to them, "Repent, and each of you be baptized in the name of Jesus Christ for the forgiveness of your sins; and you will receive the gift of the Holy Spirit.39"For the promise is for you and your children and for all who are far off, as many as the Lord our God will call to Himself."... Amen.

https://www.facebook.com/The.Illogical.Atheist/photos/a.355584184549936.1073741826.339002736208081/632415430200142/?type=1&theater - the evolutionary paradigm forced upon mass millions today is provably religious in nature and the IMPOSSIBLE mental hurdles such people espouse not only rivals all science and rational thought, BUT IF IT WERE TRUE would be NO LESS MIRACULOUS than the Biblical Account of Creation. The supernatural miracles or rather the IMPOSSIBILITIES atheists and evolutionists believe in the so-called name of science:

https://www.facebook.com/photo.php?fbid=982807705071177&set=gm.575896999217802&type=1&t heater

http://www.blastthetrumpet.org/PublicLetters/AAAUpdatedPublicAlertsMattersofLifeandDeath/More Updates041215/The%20religion%20of%20evolution%20versus%20the%20Science%20of%20Creation.pd f) as you can read, I take a bit stronger position in asserting CONTRARY to evolution WHAT WE DO OBSERVE is mass extinctions, EVEN mass extinctions today, disease and death from genetic LOSS of once healthy specimens over time! (the EXACT OPPOSITE of evolutionary claims) in other words REALITY, the fossil record if anything puts the evolutionary tree of life UPSIDE DOWN because it plainly shows enormous numbers of complex multicelled species that are DYING OFF over time with less and less species remaining on earth! and THE ONES STILL ALIVE, even OURSELVES, are LOSING genetic information causing birth defects and making even ourselves as a species sicker and sicker while we watch extinction events worldwide taking place

https://www.youtube.com/results?search_query=mass+die+offs+2014+extinction+events and

https://www.youtube.com/watch?v=lf9yzHwOeUc and

https://www.blastthetrumpet.org/PublicLetters/AAAUpdatedPublicAlertsMattersofLifeandDeath/Updates053016/End%20Times%20Verse%20By%20Verse.pdf so while delusionary evolutionists harp on about their wacky ideas, EVERYTHING we OBSERVE shows it to be FALSE and DIAMETRICALLY opposed to their foolish notions.

http://www.uncommondescent.com/intelligent-design/a-world-famous-chemist-tells-the-truththeres-no-scientist-alive-today-who-understands-macroevolution/

http://creation.mobi/y-chromosome-shock

https://www.facebook.com/156650811060818/photos/o.514985451851831/973161102743114/?typ e=1&theater This is the scope of the complete load of crap (the whole gamut of the modern evolutionary paradigm); fiction in the name of science, being forced upon innocent children in public brainwashing https://www.blastthetrumpet.org/PublicLetters/AAAUpdatedPublicAlertsMattersofLifeandDeath/Updates053016/NO%20MORE%20BRAINWASHING%20p1.pdf and https://www.blastthetrumpet.org/PublicLetters/AAAUpdatedPublicAlertsMattersofLifeandDeath/Updates053016/NO%20MORE%20BRAINWASHING%20p2.pdf; indoctrination programs in America and all over the world:

1) EVERYTHING came from nothing - https://www.google.com/webhp?sourceid=chromeinstant&rlz=1C1GIGM_enUS535US535&ion=1&espv=2&ie=UTF8#q=the%20universe%20came%20from%20nothing%20according%20to%20some%20scientists

It doesn't matter what their excuse for such mental acrobatics and nonsensical imaginations are, when ANYONE VIOLATES THE FOUNDATIONAL LAWS OF KNOWLEDGE AND SCIENCE itself to believe something so OBVIOUSLY FALSE, they have passed the state of willful ignorance, stupidity, and are well on their way to total insanity! - The "universe from nothing" violates the Laws of Non-contradiction and Causality; which as Aristotle put it, must exist for anyone to know anything at all. (Evolutionary Cosmology, as well as Evolution, is being taught in the name of science today is OVERTLY RELIGIOUS; not even remotely rationally supportable, and is MOST DEFINITELY NOT scientific.)

https://www.youtube.com/watch?v=v34QjYPuiEA -

https://www.blastthetrumpet.org/PublicLetters/AAAUpdatedPublicAlertsMattersofLifeandDeath/Updates053016/The%20Dawkins%20Delusion.pdf

2) They go on to claim the Causeless EVERYTHING universe not only came from nothing, BUT that includes ALL forms of energy and matter! (Chemical evolution of how atoms formed, molecules, and all the various elements and forms of energy, have yet to be demonstrated by anyone and as such the "theories" fall once again outside the realm of science and the scientific method and are purely religious). YET is all being taught as FACT even though those teaching such things cannot demonstrate what they BELIEVE by blind faith. (everything causeless came from nothing and without directed cause mysteriously assembled itself into separate forms of energy and elements).

3) Atheists and evolutionists further go on to espouse their religious beliefs regarding stellar evolution. How not only the causeless universe that came from nothing causelessly assembled itself into all known forms of energy and matter, BUT that those forms of energy and matter were so balanced as to combine into stars, planets, asteroids, comets, etc. and even our earth. WHY are there larger galaxies, smaller galaxies, larger stars, smaller stars, etc. IF it all began with ONE form of energy and undefined massless particles, HOW did assembly not only atomically occur, the elements, but VARIED concentrations of elements in stars and planets? AND why did assembly occur at all and HOW are these various masses of various elements and properties held together; in clear violation of the Law of Entropy and the scientific properties of those various elements and isotopes? With every step one takes in the modern popular evolutionary paradigm, one gets further and further away from TRUTH and reality into the depths of abject delusions, deceptions and irrational fantasies!

4) Abiogenesis - the nonsense being forced on mass millions in the name of so-called science, DISREGARDS SCIENCE and WHAT WE ALL OBSERVE to make ANOTHER IMPOSSIBLE CLAIM! The Law of Biogenesis alone is IRREFUTABLE EVIDENCE of our ETERNAL CREATOR! LIFE COMES FROM LIFE -

https://www.blastthetrumpet.org/PublicLetters/AAAUpdatedPublicAlertsMattersofLifeandDeath/Updates053016/GOD%20Part%201.pdf and https://www.blastthetrumpet.org/PublicLetters/AAAUpdatedPublicAlertsMattersofLifeandDeath/Updates053016/GOD%20Part%202.pdf - it absolutely does NOT come from the nothing that became everything that causelessly defined itself into elements and forms of energy, and to top it all off those causeless, chaotic particles and energy not only assembled themselves together but without any reason whatsoever SPRANG TO LIFE one day!!!!!!!! WHAT IS BEING FORCED ON CHILDREN as the evolutionary constructs taught in the name of science today IS TOTAL MADNESS!!!!!!!! http://biblehub.com/romans/1-22.htm and http://biblehub.com/romans/1- 28.htm

5) SO FAR, the IMPOSSIBLE hurdles that atheists and evolutionists mentally jump are: a) everything came from nothing b) forms of energy and defined atomic structure "evolved" into existence without any rational explanation as to why or how c) in defiance of entropy and elemental properties these invisible particulates assembled themselves into visible matter of galaxies, stars, planets in various sizes and dimensions, in various concentrations from the causeless "cosmic inflation" event. d) life sprang into existence from non-life. **ALL OF THESE MENTAL GYMNASTICS VIOLATE THE LAWS OF SCIENCE, KNOWLEDGE AND REASON!!!!!!!! everything from**

nothing, order from chaos, life from non-life are ALL DIRECT CONTRADICTIONS! otherwise known as IMPOSSIBILITIES!!!!!!!! PLEASE DO NOT IGNORE THE COMMENTS SECTIONS OF ANY OF MY NOTES; WHICH ARE LOADED WITH CITATIONS OF FACTS SUPPORTING ALL MY STATEMENTS AND CONCLUSIONS. So, you get people on earth teaching little kids that all that is necessary for life to exist, is earth, water, air or atmosphere, and sun or similar energy. BUT IN ALL THE PRESENTATIONS I HAVE SEEN THEY ALL LEAVE OFF THE MOST CRITICAL OF ALL ATRIBUTES NECESSARY FOR LIFE - INFORMATION! Without DNA DEFINED INSTRUCTIONS, the elements will NEVER "come to life" NO MATTER HOW MUCH "time" elapses! LIFE is MUCH MORE than just hardware, (elements and energy), IT IS DEFINED BY DNA INFORMATION, INFORMATION SO COMPLEX that super computers are STILL analyzing it! And the sad truth is these genomes are only the beginning of finding out just HOW DETAILED. WITHOUT BOTH HARDWARE (elements, energy) and SOFTWARE (intelligence, knowledge, information) you will never have a working machine, you will never get ANYTHING that even remotely resembles LIFE! THE VERY PRESENCE OF BOTH HARDWARE (the physical universe) AND SOFTWARE (information/knowledge/science) REQUIRES ENGINEERING; IS ABSOLUTELY IRREFUTABLE PROOF OF INTELLIGENT DESIGN BY OUR ETERNAL CREATOR!!!!!!!! When I have pointed out to atheists and evolutionists these clear impossibilities, and how utterly foolish it is to believe this nonsense, they CLAIM they are not teaching these things to students in public primary, secondary educational institutions (coerced indoctrination) and even at collegiate levels; BUT THEY ARE!

https://www.youtube.com/watch?v=3HoHmSn5Gns and then they reinforce the public brainwashing in media.

https://www.blastthetrumpet.org/PublicLetters/AAAUpdatedPublicAlertsMattersofLifeandDeath/Updates053016/PROPAGANDA.pdf and

http://themindunleashed.org/2014/07/ten-popularmind-control-techniques-used-today.html and

https://www.youtube.com/watch?v=VllwRgSECcw and as social engineering demoralization expert exposed

https://www.youtube.com/watch?v=kJKJM36lMTE

Recapping that cosmic evolution as taught today is FALSE and IMPOSSIBLE, chemical evolution as it is taught today is FALSE and IMPOSSIBLE, stellar evolution as it is taught today is FALSE and IMPOSSIBLE, organic evolution as it is taught today is FALSE and IMPOSSIBLE; brings us to the fifth (PROMINENT impossibility; there are innumerable impossibilities atheists and evolutionists have to believe in order to hold to their delusionary worldviews) OVERTLY OBVIOUS IMPOSSIBILITY of macro evolution from a single cell into the billions of more complex species that have ever existed. THERE ARE SO MANY REASONS WHY macro-evolution is IMPOSSIBLE that it's nearly impossible to list them all.

https://www.youtube.com/watch?v=GjvuwneORrE BUT AT THE CORE of macro-evolutionary IMPOSSIBILITY is the belief that GREATER COMPLEXITY comes from simplicity, GREATER DETAILED KNOWLEDGE comes from ignorance, and that myriads of diverse multi-celled organisms all "evolved" from a single celled organism; including our multi-trillion celled selves. If there is any evidence at all for the modern evolutionary paradigm, it is the remaining pond scum left in the heads of evolutionists and atheists for what should be a brain, because that is how MONUMENTALLY IGNORANT one has to be to espouse this total rubbish and madness! -

https://www.blastthetrumpet.org/PublicLetters/AAAUpdatedPublicAlertsMattersofLifeandDeath/Updates053016/Evolution%20and%20Atheism%20Intertwined%20Cults%20of%20the%20Insane.pdf - THIS IS WHY THE ORIGINAL MEME cited states the TRUTH that what atheists and evolutionists espouse in the name of science is NOT ONLY IMPOSSIBLE, but if ANY OF IT were remotely plausible, would be no less SUPERNATURAL than what is recorded in the Biblical Account of Creation.
https://www.youtube.com/watch?v=_LTaPIK7maY&feature=youtu.be I want people to understand WHY persons like myself take this subject SO SERIOUSLY. We are NOT trying

to participate in exercises of intellectual debate! We are NOT trying to insult, denigrate, slander, libel, anyone on earth! We are NOT trying to garner attention for ourselves! WE ARE TRYING TO SAVE LIVES! LITERALLY SAVE LIVES! NOT JUST FREEDOMS, LIBERTIES, AND CRITICAL THOUGHT, BUT LIVES AND EVERLASTING DESTINIES!!!!!!!! https://www.youtube.com/watch?v=QslXG6YhnFY

If you are thinking we are making way too big a deal out of debating what is being taught to mass millions of souls on earth; you need to FACE THE FACTS as to WHY persons like Dr. James Kennedy, Dr. Kent Hovind, and the many who have risked their careers and even their lives in standing against this evil ideology of state indoctrination all over the world. The FACT is evolutionary constructs, as they are forced on innocent children today, played a key role in the holocaust of WWII.

https://www.youtube.com/watch?v=4mxXICZ9mXo&list=PL7F9B57EBDCCEECF8

https://www.google.com/webhp?sourceid=chrome-instant&ion=1&espv=2&ie=UTF8#q=Hitler's%20own%20words%20reveal%20how%20evolution%20was%20responsible%20for%20his% 20views%20and%20the%20holocaust

https://www.google.com/webhp?sourceid=chrome-instant&ion=1&espv=2&ie=UTF8#q=How%20eugenic%20fascism%20in%20America%20helped%20Hitler%20and%20evolution It is being used in America to impose the SAME kind of atrocities! Events are transpiring that exactly mimic what took place in Nazi Germany before the holocaust began in earnest!

https://www.google.com/webhp?sourceid=chrome-instant&ion=1&espv=2&ie=UTF8#q=how%20America%20helped%20the%20Nazis%20including%20Hitler%20to%20escape
http://www2.gwu.edu/~nsarchiv/NSAEBB/NSAEBB146/

https://www.google.com/search?q=project+paperclip+nazis+in+america&rlz=1C1GIGM_enUS535US535&oq=project+paperclip+nazis+in+america&aqs=chrome..69i57&sourceid=chrome&es_sm=93&ie=UTF-8
CONCURRENTLY WITH THE ASSIMILATION OF NAZI WAR CRIMINALS IN AMERICA AND THE AMERICAS, evolutionary brainwashing began in EARNEST in American public education! (The Nazis weren't defeated, they merely immigrated into our government, where they have usurped authority over the very souls that bled and died to stop them!)

https://www.google.com/webhp?sourceid=chrome-instant&ion=1&espv=2&ie=UTF8#q=photos%20of%20Hitler%20prove%20he%20escaped

https://www.google.com/webhp?sourceid=chrome-instant&ion=1&espv=2&ie=UTF8#q=photos+of+Nazi+war+criminals+that+escaped+to+the+Americas&tbm=vid

https://www.google.com/search?q=nazis+living+in+america&rlz=1C1GIGM_enUS535US535&oq=nazis+living+in+america&aqs=chrome..69i57&sourceid=chrome&es_sm=93&ie=UTF-8

https://www.youtube.com/watch?v=t0bOm5c43js Nazi human experimentation has been ongoing right here in America as a result.

https://www.google.com/webhp?sourceid=chromeinstant&rlz=1C1GIGM_enUS535US535&ion=1&espv=2&ie=UTF8#q=nazi%20scientists%20in%20america

https://www.google.com/search?q=nazi+human+experimentation+continued+on+in+america&rlz=1C1GIGM_enUS535US535&oq=nazi+human+experimentation

+continued+on+in+america&aqs=chrome..69i57&sourceid=chrome&es_sm=93&ie=UTF-8

https://www.google.com/search?q=American+atrocities+in+psychiatric+institutions&rlz=1C1GIGM_enUS535US535&oq=American+atrocities+in+psychiatric+institutions&aqs=chrome..69i57&sourceid=chrome&es_sm=93&ie=UTF-8

https://www.youtube.com/watch?v=_5dSZnbugpc Eugenic fascism, VERY WICKED AND EVIL PEOPLE ARE TO BLAME for the mass brainwashing in public education and media currently underway against the public, mass millions of innocent children!

https://www.google.com/search?q=how+mass+brainwashing+techniques+were+developed+and+implemented&rlz=1C1GIGM_enUS535US535&oq=how+mass+brainwashing+techniques+were+developed+and+implemented&aqs=chrome..69i57&sourceid=chrome&es_sm=93&ie=UTF-8

https://www.youtube.com/results?search_query=CIA+involvement+in+mass+brainwashing Sadly, brainwashed victims, often don't realize how they have been deceived and manipulated until it is too late; so, they don't even realize what they are doing when they say ridiculous things like why do you use google and youtube for your citations; as if to suggest the Internet as a whole has no reliable reference material stored globally. I stopped citing constantly specific references because important websites kept disappearing or became "unavailable" like: http://naturalskies.org/references which had scientific and even government documents showing their hand in the HAARP programs, geoengineering, weather warfare and mass mind control. -

https://www.blastthetrumpet.org/PublicLetters/AAAUpdatedPublicAlertsMattersofLifeandDeath/Updates053016/Crimes%20Against%20Humanity%20Part%201.pdf and

https://www.blastthetrumpet.org/PublicLetters/AAAUpdatedPublicAlertsMattersofLifeandDeath/Updates053016/Crimes%20Against%20Humanity%20Part%202.pdf - I use the worldwide database and cite such references so you can SEE for yourselves the INFORMATION and disinformation tactics in play! (Intentionally overloading citizens with so much nonsense that it becomes difficult for anyone to identify what is true and reliable and so they typically and predictively resort back to whatever they learned or were told as children and young adults and dismiss even evidence before their very eyes that disproves what they THINK is true. - https://www.youtube.com/watch?v=kJKJM36lMTE#t=105

4) "the demoralization process (mass brainwashing) in the United States is basically completed already...actually it's over fulfilled...not even comrade Andropov and all his experts would even dream of such a tremendous success! ... thanks to lack of moral standards... (https://www.google.com/search?q=comrade+andropov&rlz=1C1GIGM_enUS535US535&oq=comrade+andropov&aqs=chrome..69i57&sourceid=chrome&es_sm=93&ie=UTF-8) ... as I mentioned before, exposure to true information does not matter anymore. A person who is demoralized (brainwashed) is unable to assess true information; the facts tell nothing to him... even if I shower him with information... he will REFUSE to believe it..." DO YOU UNDERSTAND YET THE SERIOUSNESS OF THE SITUATION? America has factually and provably been infiltrated at the highest levels by ENEMIES who have taken control of our government, our military, our schools, our media, and are setting the stage for another holocaust! concentration camps! the arrests and deaths of dissidents! IT IS ALREADY BEGINNING! My notes expose WHO these enemies of America are and what they are doing and planning! PLEASE BECOME AWARE OF THESE FACTS IMMEDIATELY!

http://www.blastthetrumpet.org/PublicLetters/AAAUpdatedPublicAlertsMattersofLifeandDeath/THE%20PLANNED%20DEMORALIZATION%20OF%20AMERICA.pdf

They have been intentionally brainwashing children with the EXACT SAME knowledge and techniques that lead to the holocaust in Germany and the mass millions slain in Russia! (state coerced indoctrination of evolutionary and atheistic constructs mixed with marxist' arrogant disdain for TRUE knowledge (Intelligent Design, our Christian American Heritage; that defeated the satanists, these wicked persons; who for generations, have been intentionally trying to enslave and rule the world)

https://www.blastthetrumpet.org/PublicLetters/AAAUpdatedPublicAlertsMattersofLifeandDeath/Updates053016/Strategies%20of%20War.pdf - strategies of war

https://www.blastthetrumpet.org/PublicLetters/AAAUpdatedPublicAlertsMattersofLifeandDeath/Updates053016/Conspiracy%20Theories%20or%20Plain%20Truth.pdf - conspiracy theories or plain truth

https://www.blastthetrumpet.org/PublicLetters/AAAUpdatedPublicAlertsMattersofLifeandDeath/Updates053016/Most%20Wanted%20Criminals%20part%203.pdf - the most wanted criminals aren't necessarily the ones you're thinking of https://www.blastthetrumpet.org/PublicLetters/AAAUpdatedPublicAlertsMattersofLifeandDeath/satanism.pdf - when you delude yourself with any lie (theory of evolution) you are committing idolatry and blasphemy against the One True God. God doesn't lie! People do!

When you believe lies and pass them on it make you a liar. satanil is the father of all lies in the Holy Bible (Jn 8:43-45) so believing in evolution is satanic.

https://www.blastthetrumpet.org/PublicLetters/AAAUpdatedPublicAlertsMattersofLifeandDeath/Updates053016/AntiChrist.pdf - anti-christs are just a word for those who resist God and His Gospel Message, His Words in the Holy Bible.

Beware of Antichrists

...**21**I have not written to you because you lack knowledge of the truth, but because you have it, and because no lie comes from the truth. **22**Who is the liar, if it is not the one who denies that Jesus is the Christ? This is the antichrist, who denies the Father and the Son. **23**Whoever denies the Son does not have the Father, but whoever confesses the Son has the Father as well....
https://biblehub.com/1_john/2-22.htm

So, the lie of the Theory of Evolution is anti-christian.

https://www.facebook.com/FFAF.International/photos/a.392139044163477.91994.390996674277714/840248046019239/?type=1&theater - a side note here for people who claim Hitler was a Christian (think about how many people all around you who CLAIM to be "Christians" who are NOTHING like Christ! -

http://www.blastthetrumpet.org/PublicLetters/AAAUpdatedPublicAlertsMattersofLifeandDeath/Are%20You%20A%20Christia1.pdf) - Are you a Christian?

https://www.google.com/search?q=hitler+evolution+quotes&rlz=1C1GIG M_enUS535US535&espv=2&biw=1066&bih=702&source=lnms&tbm=isch&sa=X&ei=jQJIVJDZDsmoyASY tICoBA&ved=0CAYQ_AUoAQ&dpr=1.5 the satanic powers ALL put on a public facade (political double talk, they will say or do anything to garner the favor of the public in order to achieve their secret goals); so they like most all politicians put on a show for the public, while behind the scenes they reveal their TRUE intentions (actions speak louder than words; when you research what Hitler stated publicly against what Hitler stated privately you SEE

CLEARLY, he was practicing intentional deception by FRAMING the Messiah (Yahoshuah, aka Jesus Christ) for his evil acts while he slaughtered and imprisoned both Jews and Christians. (he was NOT a Christian; but a satanic deceiver, funded by the same satanic bloodlines STILL inciting wars and deceiving the masses today) -

https://www.google.com/webhp?sourceid=chromeinstant&rlz=1C1GIGM_enUS535US535&ion=1&espv=2&ie=UTF8#q=how%20the%20rothschilds%20funded%20hitler and

https://www.google.com/search?q=how+bush+was+connected+to+the+nazis&rlz=1C1GIGM_enUS535US535&oq=how+bush+was+connected+to+the+nazis&aqs=chrome..69i57&sourceid=chrome&es_sm=93&ie=UTF-8 Despite all the naysayers (religious supporters of the insanity of Darwinism), Hitler and the Nazi's embraced Darwin's evolutionary nonsense as scientific justification for his atrocities; which included arresting both Jews and Christians:

http://en.wikipedia.org/wiki/On_the_Origin_of_Species#mediaviewer/File:Origin_of_Species_title_pag e.jpg - notice the full title incites racism at the onset. http://creation.com/hitlers-master-race-children-hauntedby-their-past Deceptive historians try to brush off the fact of Darwinism's contribution to racism.

https://www.google.com/webhp?sourceid=chromeinstant&rlz=1C1GIGM_enUS535US535&ion=1&espv=2&ie=UTF8#q=darwin+nazi+eugenics+and+selective+breeding and the fact that key members of Hitler's Nazi party had very powerful and influential rolls in that regard.

https://www.google.com/webhp?sourceid=chromeinstant&rlz=1C1GIGM_enUS535US535&ion=1&espv=2&ie=UTF8#q=darwin+nazi+eugenics+and+selective+breeding+racism "In his book The Scientific Origins of National Socialism (1971), Daniel Gasman claimed: "Haeckel . . . was largely responsible for forging the bonds between academic science and racism in Germany in the later decades of the nineteenth century."8In a later book, Gasman urged that Haeckel had virtually begun the work of the Nazis: "For Haeckel, the Jews were the original source of the decadence and morbidity of the modern world and he sought their immediate exclusion from contemporary life and society."9 Gasman's judgment received the imprimatur of Stephen Jay Gould, who concluded in his Ontogeny and Phylogeny (1977): But as Gasman argues, Haeckel's greatest influence was, ultimately, in another tragic direction—National Socialism. His evolutionary racism; his call to the German people for racial purity and unflinching devotion to a “just” state; his belief that harsh, inexorable laws of evolution ruled human civilization and nature alike, conferring upon favored races the right to dominate others; the irrational mysticism that had always stood in strange communion with his grave words about objective science—all contributed to the rise of Nazism.10 Scholars like Gasman, Gould, Peter Bowler, Larry Arnhart—as well as a host of others—attempt to distinguish Haeckel’s views from Darwin’s, so as to exonerate the latter while sacrificing the former to the presumption of a strong causal connection with Hitler’s anti-Semitism.11 I don’t believe this effort to disengage Darwin from Haeckel can be easily accomplished, since on central matters—descent of species, struggle for existence, natural selection, inheritance of acquire characters, recapitulation theory, progressivism, hierarchy of races—no essential differences between master and disciple exist.12 So if Hitler endorsed Haeckel’s evolutionary ideas, he thereby also endorsed Darwin’s." - http://home.uchicago.edu/~rjr6/articles/Was%20Hitler%20a%20Darwinian.pdf

https://www.google.com/webhp?sourceid=chromeinstant&rlz=1C1GIGM_enUS535US535&ion=1&espv=2&ie=UTF8#q=Corrie%20ten%20boom%20hiding%20place - Hitler was a mass murderer, NOT a Christian! He murdered them! Christ tells us all such murderers are devils! http://biblehub.com/john/8-44.htm THEIR

BRAINWASHING (DEMORALIZATION - https://www.youtube.com/watch?v=n4ZxO5I-s4E) TECHNIQUES ARE WHAT IS BRINGING ABOUT SOCIALISTIC AGENDAS, PERVERSE AGENDAS, ANTICHRISTIAN AGENDAS IN THE NAME OF LIBERALISM AND "PROGRESS" IN AMERICA! THE BRAINWASHED PUBLIC THEN RECEIVES SCIENTIFICALLY STUDIED MASS MIND CONTROL TECHNIQUES THROUGH MAIN STREAM MEDIA! AND THE LATEST SECRET MILITARY (SATANIC NWO) TECHNOLOGIES!

https://www.google.com/webhp?sourceid=chromeinstant&rlz=1C1GIGM_enUS535US535&ion=1&espv=2&ie=UTF8#q=alan%20watt%20how%20television%20is%20use%20for%20mass%20brainwashing

https://www.google.com/webhp?sourceid=chromeinstant&rlz=1C1GIGM_enUS535US535&ion=1&espv=2&ie=UTF8#q=television%20scientifically%20designed%20for%20mind%20control

https://www.blastthetrumpet.org/PublicLetters/AAAUpdatedPublicAlertsMattersofLifeandDeath/Updates053016/PROPAGANDA.pdf - LOOK AT THE FACTS! EXAMINE WHO OWNS AND CONTROLS THE "BROADCAST PROGRAMMING" (synonymous with "mass brainwashing").

https://www.youtube.com/watch?v=VIlwRgSECcw Look! what they are doing to shape our culture! "Leave it to Beaver", "Andy Griffith", "Father Knows Best", etc. has been replaced with "Modern Family" and other such perverse programming like "How to Get Away With Murder" (not trying to insult anyone, just pointing out that those who currently FUND public programming have a radically different agenda than what America was founded on and what has made America such a great nation!) INTENTIONALLY TO REINFORCE https://www.youtube.com/results?search_query=hooking+kids+on+sex in forced public indoctrination. HORMONES ARE IN PUBLIC DRINKING WATER! NOT JUST TOXIC CHEMICALS!

https://www.google.com/webhp?sourceid=chromeinstant&rlz=1C1GIGM_enUS535US535&ion=1&espv=2&ie=UTF8#q=HORMONES%20HAVE%20BEEN%20FOUND%20IN%20PUBLIC%20DRINKING%20WATER%20SYSTEM S

It's NO WONDER some kids don't know if they're a boy or girl any longer! Between fluoride (https://www.google.com/webhp?sourceid=chromeinstant&rlz=1C1GIGM_enUS535US535&ion=1&espv=2&ie=UTF-8#q=fluoride%20neurotoxin), hormones and toxins in our water, poisons in our food and practically everything we drink, ingest, put on our skin:

https://www.facebook.com/notes/michael-swenson/depopulation-by-intent-oraccident/544721958940101 AND EVEN THE AIR WE BREATHE

https://www.google.com/webhp?sourceid=chromeinstant&rlz=1C1GIGM_enUS535US535&ion=1&espv=2&ie=UTF8#q=chemtrails%20haarp%20and%20mind%20control, it's a PURE WONDER that the entire world hasn't been turned into sick and dying, brain-dead persons who can't tell if they are male or female, or human or animal, or right from wrong already! Or has it?

https://www.blastthetrumpet.org/PublicLetters/AAAUpdatedPublicAlertsMattersofLifeandDeath/Updates053016/Why%20Care.pdf IF YOU THINK THAT I AM BEING EXTREME IN CALLING FOR THE IMMEDIATE GLOBAL ARRESTS OF ALL INVOLVED AND THE IMMEDIATE CESSATION OF ALL THE LIES BEING TAUGHT CHILDREN IN THE NAME OF SCIENCE THESE DAYS, THEN YOU ARE NOT LEARNING FROM HISTORY AND ARE IN DENIAL THAT AMERICA IS ALREADY TAKING VERY DANGEROUS STEPS TOWARD ANOTHER HOLOCAUST!!!!!!!! LOOK at what is ALREADY underway!

https://www.blastthetrumpet.org/PublicLetters/AAAUpdatedPublicAlertsMattersofLifeandDeath/Updates053016/Depopulation.pdf and https://www.blastthetrumpet.org/PublicLetters/AAAUpda

tedPublicAlertsMattersofLifeandDeath/Updates053016/Strategies%20of%20War.pdf (Besides the brainwashing, poisons in food, air, water, hygiene products, medications, besides the nazis, socialists, in education and government, besides the overt demoralization taking place in schools and media actively destroying our American Christian Heritage and highest laws and liberties thereby, besides enemy infiltration at highest levels of government and within our borders (https://www.blastthetrumpet.org/PublicLetters/AAAUpdatedPublicAlertsMattersofLifeandDeath/Updates053016/What%20is%20Happening%20to%20America.pdf), besides the latest technologies and military tactics being used against us all to enslave us

https://www.blastthetrumpet.org/PublicLetters/AAAUpdatedPublicAlertsMattersofLifeandDeath/Updates053016/Police%20State%20Prison%20State%20Big%20Business%20Modern%20Slavery.pdf and all due to the personal evil, greed and corruption at the highest levels selling us all out and oppressing us

https://www.blastthetrumpet.org/PublicLetters/AAAUpdatedPublicAlertsMattersofLifeandDeath/Updates053016/Treatise%20on%20Greed%20and%20Corruption.pdf IT IS ABSOLUTELY CLEAR THAT THESE WICKED IDEOLOGIES (fictional evolutionary constructs combined with communism, socialism, fascism, marxism ALL state FORCED INDOCTRINATION) is ALREADY FOLLOWING THE EXACT STEPS THAT LED TO MASS MILLIONS OF DEATHS IN GERMANY, RUSSIA, CHINA (the other members of the UN "security council") PEOPLE! WAKE UP!!!!!!!!

https://www.google.com/webhp?sourceid=chromeinstant&rlz=1C1GIGM_enUS535US535&ion=1&espv=2&ie=UTF8#q=the%20homeless%20are%20being%20rounded%20up%20in%20camps - concentration camps have been constructed and are being manned for full operation!

https://www.youtube.com/results?search_query=jesse+ventura+fema+camps+

https://www.google.com/webhp?sourceid=chromeinstant&rlz=1C1GIGM_enUS535US535&ion=1&espv=2&ie=UTF8#q=creationists%20intellectual%20educators%20are%20being%20fired%20across%20America - the Nazis and Communists ALSO FIRED and ARRESTED the educators; to replace them with their state sponsored indoctrinators!

https://www.intellihub.com/scientist-terminated-proving-dinosaurs-humans-walked-earth-together/ Dr. Kent Hovind was asked by a naive individual why he was facing off against this evil brainwashing; a young student who simply couldn't comprehend the fact that Christians are already fighting for not just our freedoms but our lives as a result!

https://www.youtube.com/watch?v=ED32UZT06sc Dr. Hovind is still in prison today! (When I first started writing this book; but was subsequently set free from his incarceration.) But I have noticed his videos disappearing from youtube!
https://www.google.com/webhp?sourceid=chromeinstant&rlz=1C1GIGM_enUS535US535&ion=1&espv=2&ie=UTF-8#q=Kent+Hovind+still+in+prison and http://www.pnj.com/story/news/2014/11/08/dr-dino-facing-new-legal-woes/18730293/ this is overt persecution by satanists wielding their greedy arm of perverted so called justice. The federal reserve and the oppressive arm they control the IRS has imposed tribute to themselves in the form of taxing income, a right not granted or afforded them at all and in fact Americans fought for independence from such overt oppression by the wealthy against the poor. The Constitution as our nation was founded DOES NOT AUTHORIZE these wicked persons in the Federal Reserve and IRS oppressors to operate! It is only by infiltration and acts of treason that these criminal organizations exist at all! And so, the satanic families wield their might by oppressing the masses and especially those who expose their criminal activities like Kent Hovind has

done. (subjugating the masses by brainwashing them in forced public indoctrination), and creating fictitious "laws" that are factually unenforceable under our Constitution, but until Americans find the desire for freedom that our founding fathers had, these wicked people will continue to get away with their evil oppression and arbitrary "justice" that they selectively use to silence any and all opponents in a so-called free civilization.

https://www.facebook.com/notes/heather-ritz/scientists-and-professors-who-have-facedostracization-persecution-andor-discrim/758604884184284 - people losing their careers over this issue.

http://www.naturalnews.com/046315_dinosaur_fossils_soft_tissue_wrongful_termination.html PUT THE SHOE ON THE OTHER FOOT! HOW WOULD evolutionists FEEL IF THEY WERE THE ONES LOSING THEIR CAREERS; THAT THEY HAD SPENT ALL THEIR EARNINGS AND LIFE TO DEVELOP SUCH SKILLS FOR, THEIR ABILITY TO EARN A LIVING, THEIR ABILITY TO SURVIVE, AND WERE ACTIVELY BEING PERSECUTED SIMPLY FOR WHAT THEY BELIEVE AS TRUE KNOWLEDGE! LOOK AT HOW MANY EDUCATORS AND SCIENTISTS HAVE LOST THEIR JOBS OVER THIS! IT IS VERY SERIOUS!

https://www.google.com/webhp?sourceid=chromeinstant&rlz=1C1GIGM_enUS535US535&ion=1&espv=2&ie=UTF8#q=SCIENTISTS+ARE+LOSING+THEIR+JOBS+OVER+CREATIONISM - not just educators but other professionals are being fired over this issue!

https://www.google.com/webhp?sourceid=chromeinstant&rlz=1C1GIGM_enUS535US535&ion=1&espv=2&ie=UTF8#q=professionals+fired+in+all+kinds+of+jobs+over+creationism BUT it is worse than this, Christians are being sued for simply adhering to our beliefs and some are ending up in jail or prison over them ALREADY! (this is exactly how the previous holocaust began)
https://www.google.com/webhp?sourceid=chromeinstant&rlz=1C1GIGM_enUS535US535&ion=1&espv=2&ie=UTF8#q=Preachers%20in%20America%20are%20going%20to%20jail%20because%20of%20liberal%20homosexuals - HOW DID WE GO FROM SODOMY LAWS AND THE RIGHTEOUS MINDSET OF OUR FOUNDING FATHERS, TO HAVING PEOPLE THAT COMPLETELY IGNORE THE HIGHEST LAWS OF OUR LAND?
http://sonsoflibertymedia.com/2014/10/founding-fathers-believed-homosexuality/ you guessed it! BRAINWASHING in schools and media! CORRUPTION in the highest levels!

https://www.youtube.com/watch?v=kJKJM36lMTE#t=1031

https://www.blastthetrumpet.org/PublicLetters/AAAUpdatedPublicAlertsMattersofLifeandDeath/Updates053016/Americans%20and%20Citizens%20of%20the%20World%20FREEDOM%20and%20LIFE%20Comes%20from%20GOD.pdf BUT you STILL say I am taking this too far to say evolutionary brainwashing in public education is to blame for the rise of concentration camps in America; complete with rail cars fitted with shackles and tracks that lead straight to the camps now being manned and operated.

https://www.google.com/webhp?sourceid=chromeinstant&rlz=1C1GIGM_enUS535US535&ion=1&espv=2&ie=UTF8#q=passenger%20cars%20with%20shackles%20in%20America%20lead%20straight%20to%20fema%20camps - fema camps and fusion centers

https://www.youtube.com/results?search_query=passenger+cars+with+shackles+in+America+lead+straight+to+fema+camps - what are they doing?

https://www.youtube.com/results?search_query=fusion+centers+exposed - fusion centers on American soil

https://www.google.com/webhp?sourceid=chromeinstant&rlz=1C1GIGM_enUS535US535&ion=1&espv=2&ie=UTF8#q=how%20evolution%20led%20to%20the%20holocaust evolution and the holocaust

https://www.google.com/webhp?sourceid=chromeinstant&rlz=1C1GIGM_enUS535US535&ion=1&espv=2&ie=UTF-8#q=evolution%20and%20communism evolution and communism

https://www.google.com/webhp?sourceid=chromeinstant&rlz=1C1GIGM_enUS535US535&ion=1&espv=2&ie=UTF8#q=evolution+and+atheism+caused+millions+of+deaths+in+Germany%2C+Russia+and+China - the link between evolutionary thought and state sponsored indoctrination that has led to hundreds of millions of deaths and loss of liberty and human rights globally is PLAIN TO SEE CURRENTLY AND IN HISTORY! But WHEN IT IS FOLLOWING THE EXACT SAME PROCEDURES HERE IN AMERICA and people are STILL thinking all of us who are sounding the alarm are nothing more than "conspiracy theorists" or that we are making too big an issue out of this; by telling you all to arm yourselves immediately, SUE ANY AND EVERY TIME the state or anyone steps on your liberties, and FIGHT BACK against these most wicked and SCIENTIFICALLY INSUPPORTABLE worldviews BY UNITING EVERYWHERE AND DEMANDING THIS EVIL crap BE REMOVED

IMMEDIATELY FROM PUBLIC EDUCATION, DEMAND THOSE WHO HAVE BEEN FIRED OVER THIS ISSUE BE COMPESATED AND DEMAND THOSE WHO HAVE BEEN INCARCERATED BECAUSE OF THESE EVIL IDEOLOGIES, BE RELEASED IMMEDIATELY! (turning your eyes away from what is happening or remaining silent is EXACTLY what allowed mass millions of innocent persons to be slaughtered ALL OVER THE WORLD!) So, understand WHY we are standing boldly against this great wickedness and deception! WE SEE CLEARLY IT IS A MATTER OF NOT JUST FREEDOM, BUT LIFE AND DEATH! PLEASE TAKE THE TIME TO VIEW THE EVIDENCE! evolutionary theory as it is forced upon the masses today does NOT qualify as science! IT DOESN'T EVEN QUALIFY AS SCIENCE FICTION! AND IT IS PROVABLY VERY DANGEROUS, VERY EVIL, AND VERY DESTRUCTIVE! I call for the immediate release of the Brave American Patriot and Defender of Truth, Dr. Kent Hovind IMMEDIATELY (and all like him), and I call for the arrests of all those involved in his unjust persecution!!! IT IS HIGHLY IMPROPER OF THE STATE or ANY "officials" to use their conspiratorial powers in order to single citizens out for expressing their righteous and scientifically supported beliefs; against their now vast set of laws that are so large and numerous that virtually EVERYONE has violated something in them (ENTIRE WAREHOUSES OF OPPRESSIVE AND UNJUST LAWS HAVE BEEN CREATED BY THE WICKED AND CORRUPT, BRIBE TAKING POLITICIANS; SO THEY CAN LOCK UP ANYONE AT ANY TIME; BECAUSE NO ONE CAN EVEN READ THEM ALL; LET ALONE KEEP EVERY PARTICULAR!) Evolutionists and the resulting atheists from the forced indoctrination of evolutionary constructs in public education laugh at, and sneer in derision at Creationists, claiming that Creationists do not base their worldview on evidence or reason; while they do. (They imagine themselves to be the intelligent ones; while they simultaneously imagine Creationists to be stupid, dumb, ignorant people who only have come to their conclusions by blind faith). Let's examine the FACTS in that regard. When I say brainwashing is evident in these people clearly by what they say. Things like "but we observe evolution every day!" That connotation of "evolution" that they are talking about has been distilled down to "change over time". THE ENTIRE UNIVERSE IS IN MOTION! SO OF COURSE IT IS ALL CHANGING! (with the exception of the Character of our Eternal Creator) EVERYTHING WE OBSERVE IS UNIQUE! - it's ALL different! So, to say because everything is unique and moving; thereby changing; and to then take that observation and use it as the definition to claim a bacteria morphed into billions of species; including our multi-trillion celled selves is NOT scientific; but PURE IMAGINATION. So, you ask them for examples, there are none; but they will say bacteria! bacteria have become resistant in recent history! And they go on to say things like "we have observed viruses that ADD beneficial DNA information to bacteria that make them MORE complex than they were before." Ask them, "who is 'we'?" Who do you know owns and has access to electron microscopes and biological microscopes powerful enough to OBSERVE "viruses" and OBSERVE them SO CLOSELY AND THOROUGHLY that they can SEE the virus imparting the

DNA that supposedly wasn't in the bacteria previously? This is what I mean by "brainwashed". They idolize "science" and "scientists" SO MUCH that they will REPEAT ANYTHING THEY ARE TOLD; AND EVEN CLAIM THEY SAW IT THEMSELVES by using the all inclusive "we observed"; when they did NOTHING OF THE KIND! (Are LIARS, or self-deluded, or deceived persons). But they pull that rubbish out of their wild imaginations DESPERATELY to avoid the TRUTH of Intelligent Design. Let's say that they actually have documented a DNA strand that wasn't DAMAGED by a virus, (every virus the rest of us on planet earth have encountered are BAD NEWS, make us sick, destroy working computers, etc. BUT THESE PEOPLE want ALL THE REST OF US TO TAKE THEIR WORD THAT THEY ACTUALLY SAW a virus changing DNA (the very defining information and NOT only changing it, but actually changed it into MORE working and competent COMPLEXITY). NOTHING WE OBSERVE MORPHS/EVOLVES ONE KIND OF CREATURE INTO ANOTHER KIND! It's purely cultic beliefs! Ask yourselves how many of you think all these people saying they have observed this are telling the truth? We all have access to the worldwide database and I have seen NO EVIDENCE WHATSOEVER proving this nonsense. Furthermore, then ask yourselves, that were this actually true, WHICH IS MORE REASONABLE TO BELIEVE? that viruses causing death and illnesses to organisms ubiquitously have suddenly become the cause of beneficial added information in DNA sequencing, OR that they are only DESPERATELY CLAIMING SUCH DUE TO A COMPLETE AND TOTAL LACK OF OBSERVABLE EVIDENCE for

Darwinism that they try in vain to support with ridiculous claims like this. Which is more likely? that in an era of biological engineering and genetic modification, we are seeing resistant strains, because the powers that be have been intentionally creating biological weapons OR that previously eradicated diseases have suddenly done what they never did in the past? LOOK AT THE ACTUAL EVIDENCE - biological warfare by eugenic fascists against the masses AS THEY HAVE PATENTED THEIR DESIGNS!! - https://www.google.com/webhp?sourceid=chromeinstant&rlz=1C1GIGM_enUS535US535&ion=1&espv=2&ie=UTF-8#q=patent%20of%20ebola

http://www.google.com/patents/US4647773

https://www.google.com/webhp?sourceid=chromeinstant&rlz=1C1GIGM_enUS535US535&ion=1&espv=2&ie=UTF-8#q=modern%20biological%20weapons With the advent of DNA sequencing and genetic modification on top of proven depopulation agendas, suddenly finding disease resistant strains is NOT grounds to support the nonsense of evolution, but grounds to convict the criminals attacking the general population on earth! - just because some strains of bacteria are resistant, THEY ARE STILL BACTERIA they DIDN'T MORPH/EVOLVE into anything else!

https://www.google.com/search?q=Depopulation&rlz=1C1GIGM_enUS535US535&oq=Depopulation&aqs=chrome..69i57j69i60j69i65&sourceid=chrome&es_sm=0&ie=UTF-8 - of course the propagandists are going to deny what they are doing BECAUSE IT IS OPEN GENOCIDE AGAINST HUMANITY; but LOOK AT THE OVERWHELMING EVIDENCE! -

https://www.blastthetrumpet.org/PublicLetters/AAAUpdatedPublicAlertsMattersofLifeandDeath/Updates053016/Depopulation.pdf (or would you rather watch yourself and the world die off while you desperately cling to your delusionary theories of fictional evolutionary constructs) VARIATIONS WITHIN canines, felines, bovines, etc. occurs; yes, even in bacteria. Our Creator equipped living organisms with defense mechanisms to fight off foreign invaders, including harmful viruses! In humans we have identifier cells among our leukocytes and killer cells, so we actually have DNA programming that is SO SMART it is provably SELF-LEARNING, to ward off diseases and is the real reason for the massive rise in food allergies in recent history. Our immune system is rejecting all the GMO food products in our stores now - evidence shows those allergies virtually instantaneously disappear when eating the food our Creator designed! -

https://www.google.com/webhp?sourceid=chromeinstant&rlz=1C1GIGM_enUS535US535&ion=1&espv=2&ie=UTF8#q=food%20allergies%20went%20away%20as%20soon%20as%20we%20stopped%20eating%20GMOs So when you can see the EVIL people have poisoned our food, genetically modified our food, patented diseases in order to create biological weapons of mass death and destruction, poisoning our water, air and hygiene products, inoculations and medications and STILL prefer to believe in your ridiculous evolutionary constructs of thinking the universe came from nothing and elements, energy and mass magically appeared without one shred of rational explanation, and life came from nonlife when it has been proven by the world's leading scientists to be IMPOSSIBLE: Fred Hoyle and his colleague Chandra Wickramasinghe calculated the odds that all the functional proteins necessary for life might form in one place by random events. They came up with one chance in 10 to the 40,000th power (that's 1 with 40,000 zeros after it). Since there are only about 10 to the 80th power subatomic particles in the entire visible universe, the physicists concluded that this was "an outrageously small probability that could not be faced even if the whole universe consisted of organic soup."—SHOW ME GOD, by Fred Heeren. (more quotes from the research of scientists, Nobel Prize winning scientists, widely considered as the most intelligent and powerful persons in history, discoverers of the longest lasting scientific laws, the very founders of the modern fields of scientific disciplines!:

https://www.blastthetrumpet.org/PublicLetters/AAAUpdatedPublicAlertsMattersofLifeandDeath/Updates053016/Fathers%20of%20the%20Sciences%20and%20Great%20Scientific%20Minds.pdf)

BUT even if microscopic bacteria also have a programmed defense mechanism to survive foreign invaders, toxins, etc. and thereby develop stronger immunities against such attack on those living organisms; THAT WOULD ONLY BE CONSISTENT with Intelligent Design NOT EVOLUTION! Consider the fact that our Creator warned us when we depart from Him and His Instruction then one of the curses would be that diseases would be multiplied upon us (even if the cause is genetic modification, biological weapons, the prophecy is fulfilled BECAUSE GOD merely states such things would come to pass! In other words, when people depart from Him, they become so wicked and evil that they conceive of methods of genocides) So the presence of disease, death and destruction proportionately increasing in America and the world correlates with His warning; NOT evolution! (People say I am making a mistake between cause and correlation because as I have pointed out they have already been brainwashed into a defense mechanism preventing them from acknowledging what is true.) WHEN THE ABUNDANCE OF FACTS SUPPORT THE BIBLICAL ACCOUNT and not the fictional paradigm of evolutionary imaginations OVERWHELMINGLY; then it is not I who am in error; it is those who look at the universe and all that is happening in the world and can even SEE verse by verse explanation for it all and STILL not recognize the Cause.

(they are factually brainwashed) READ ON TO SEE THE LISTS OF EMPIRICAL EVIDENCE THAT SUPPORTS THE BIBLICAL ACCOUNT!

https://www.blastthetrumpet.org/PublicLetters/AAAUpdatedPublicAlertsMattersofLifeandDeath/Updates053016/The%20Sufferings%20of%20Christ.pdf When you IGNORE REALITY AND THE MOST APPARENT AND REASONABLE EXPLANATIONS FOR WHAT WE OBSERVE, then the overwhelming EVIDENCE is that you are BRAINWASHED, delusionary, deceived, INCAPABLE of facing REALITY! (The evil brainwashers have succeeded in eliminating your ability to discern what is true or not; they have destroyed your critical thinking skills; just like the experts plainly state:

https://www.youtube.com/watch?v=kJKJM36lMTE#t=1098 They have INTENTIONALLY placed their professors in public education institutions and filed suits with corrupt judges to remove the Holy Bible and even the mention of God from public education in order to accomplish this; as I have already stated above and as history records.

https://www.google.com/webhp?sourceid=chromeinstant&rlz=1C1GIGM_enUS535US535&ion=1&espv=2&ie=UTF8#q=the%20rewriting%20of%20american%20history -

rewriting American history to leave out mention of God our Creator.

https://shop.wallbuilders.com/the-american-heritage-series-10-dvd-boxed-set - (worth every penny)

https://www.google.com/webhp?sourceid=chromeinstant&rlz=1C1GIGM_enUS535US535&ion=1&espv=2&ie=UTF8#q=The%20ACLU%20link%20to%20communism - people fighting in our nation to destroy our nation and the Freedoms Afforded us under our American Christian Heritage.

http://www.blastthetrumpet.org/PublicLetters/AAAUpdatedPublicAlertsMattersofLifeandDeath/THE%20PLANNED%20DEMORALIZATION%20OF%20AMERICA.pdf - enemy infiltrators have slipped in to our nation to demoralize Americans from our American Christian Heritage, because that is what made our nation great!

https://www.google.com/webhp?sourceid=chromeinstant&rlz=1C1GIGM_enUS535US535&ion=1&espv=2&ie=UTF-8#q=stages%20of%20demoralization Nothing about darwinian evolutionary theory or the entire evolutionary popular constructs from the origin of our universe to all their irrational explanations along the way; as are being forced upon mass millions today, are scientific in any way; other than showing how effective mass brainwashing is upon developing minds. The current fictional paradigm ("from goo to you by way of the zoo" - https://www.youtube.com/watch?v=2v5nAYU2GD0); forced upon innocent children in the name of science universally contains scientifically insupportable information. In other words, from the foundation of evolutionary constructs; throughout all the published lies, deceptions and nonsense, science shows plainly that it is all pure fiction. Science is not just the ways we pursue knowledge; it is knowledge. The Scientific Principle describes the processes or methods whereby knowledge is acquired. That body of knowledge, by the agreement of many, has become the many fields of science today. So, Science is Observable, Testable, and Repeatable; ideally, but should AT LEAST meet the OBSERVABLE criterion. Are the evolutionary teachings, the popular paradigm forced on mass millions of innocent children today; of cosmology, the origin of our universe, observable? Is it testable? Is it Repeatable? NO! NO! NO! It is therefore NOT science or scientific in any way! (Absolutely should NOT be taught as such!) How about chemical evolution as it is forced on children today? Do you know of ANYONE who has rationally explained where energy came from, how the variations of energy and matter came into existence, or has taken quarks and formed the entire list of elements in the periodic table? No? I do; our CREATOR, but not one "scientist" or the entire sum of them in all recorded history has! Therefore, WHAT IS BEING TAUGHT in the Name of Science to mass millions of innocent children regarding the origin of our universe (the explanation for what we do observe) does NOT qualify as science; NOR does the very next stage in the evolutionary constructs pass the test of "observable, testable, repeatable" in order to acquire knowledge. In other words, the first two stages of evolutionary theory both in cosmology and in any attempt to explain even the existence of elements as is widely taught DOES NOT fall within the definition of science! Moving on to stellar evolution. I meet people online these days who CLAIM to have observed the "formation of stars", the "birth of stars", even the formation of whole galaxies! (https://herschel.jpl.nasa.gov/stars.shtml) We now have access via the Internet to the largest database of information in the history of the world and yet, NOT ONE video or even time lapse series has been posted SHOWING such formation of stars and galaxies (I'm NOT talking about animations or unsubstantiated claims, I'm specifically looking for the ACTUAL PROOF that they have FACTUALLY OBSERVED stars and galaxies forming; that so many CLAIM! Even though I am fully aware of the brainwashing techniques being used on these people through public education and media, I am still shocked at how many of these young people make actual statements like "we have observed" when they observed NOTHING OF THE KIND but are regurgitating something they were taught; without question! (i.e. cosmic evolution as it is taught today does NOT meet the criterion of "science" or the scientific principle; nor does chemical evolution as it is taught today, and neither, DESPITE ALL THE HYPE OF THOSE WHO ADAMANTLY INSIST OTHERWISE, has anyone ever actually observed the formation of a star, stars, let alone one or more whole galaxies (except our Creator and to whomever He has personally shown such things to). So far, the premises, constructs, and conclusions of evolutionary theory in public indoctrination institutions today are clearly teaching FICTION, pure imaginations, in the name of science!

If any of the claims of evolution were true, we would see mass millions of slight changes in the fossil record showing it is. There is NO "natural" mechanism for "evolving" from asexual single cells that reproduce through mitosis to highly complex multicelled organisms that reproduce sexually - NONE! and yes, I've sufficiently studied "epigenetics" and all their many claims for attempting to justify something that isn't justifiable and is impossible instead. The theory of evolution is simply false.

But to continue on. We have seen that what people are being taught to think of as science is not even OBSERVED, let alone testable and repeatable! And so, we come to the NEXT HUGE IMAGINATION taught in the name of evolutionary constructs as their worldview for the "scientific" explanation of what we actually can observe. That fictional construct DEFIES what we observe by claiming life came into existence all on its own from inorganic, non-living chemicals. KEEP IN MIND, BY INTENTIONALLY OMITTING EVEN THE MENTION OF GOD, OUR ETERNAL CREATOR, WHAT THEY ARE TEACHING TO MASS MILLIONS OF INNOCENT CHILDREN; in the NAME of SCIENCE, is that

1) something came from nothing -

https://www.youtube.com/results?search_query=scientists+actually+believe+the+universe+came+fr om+nothing

2) the something that came from nothing exploded, expanded, and became time, space, energy, elements - https://www.google.com/webhp?sourceid=chromeinstant&rlz=1C1GIGM_enUS535US535&ion=1&espv=2&ie=UTF8#q=big%20bang%20today%20teaching%20how%20nothing%20that%20exploded%20became%20everything

3) the something/energy that came from nothing and that exploded/expanded to form space time; formed not only all the various forms of energy in the universe but all the various physical elements, atomic and molecular structure, that subsequently then formed all the visible matter in our universe and all the various motions of spinning stars, planets, galaxies! (the something that came from nothing

became EVERYTHING; that sure is an INCREDIBLE "nothing") but THEN we come to one of the hugest leaps of faith that these people want others to believe is science https://www.youtube.com/watch?v=Gr8Az3QQZdI

4) LIFE sprang into existence from the nothing that exploded to become everything; all on their own chemicals came together with everything that first life form needed to energize itself (live) and reproduce. I CANNOT BELIEVE ANYONE WOULD CLAIM THIS RUBBISH IS SCIENCE, but they have the audacity to laugh at anyone who questions these clearly imaginary, insupportable constructs!

http://en.wikipedia.org/wiki/Miller%E2%80%93Urey_experiment scientists are well aware of the molecular structures (chemicals) that living organisms utilize. So, to take those elements and put them in flasks, and apply what we observe heat and electrical energy will predictively yield those desired molecular structures, STILL CHEMICALS! NOT LIFE; not life by a LONG shot! Do people really think chemicals, mostly toxic chemicals is LIFE?; so NOT scientific and TOTALLY DECEPTIVE to claim "LIFE was created..." in that experiment! this experiment came NO WHERE CLOSE to creating even the most basic life form observed today!

https://www.youtube.com/watch?v=aHk_93Wn-ZU

https://www.facebook.com/edward.norkett.7/posts/903046353148580

https://s-media-cache-ak0.pinimg.com/736x/38/5a/64/385a645d0df86ae25dcab413704f87f1.jpg

5) so far ACTUAL science PROVES the ENTIRE evolutionary constructs taught today are FALSE, FICTION, PURE FANTASY, IMPOSSIBLE! NONE of the rubbish taught today regarding the reasons for the existence of the universe and thereby our own existence is scientific! It is absolute twaddle of the highest order! But then they go on to brainwash innocent children by the mass millions claiming not only all of that in the name of science; but that those chemicals that assembled themselves DESPITE EVERYTHING WE OBSERVE that contradicts those claims, that came from the magical nothing; that exploded to become everything, THEN magically transformed into ALL the BILLIONS OF SPECIES THAT HAVE EVER EXISTED ON EARTH to this day; including our multi-trillion celled selves!!!!!!!! They ACTUALLY believe that abiogenesis occurred, and they actually believe that everything came from nothing! They then furthermore ACTUALLY THINK that first life form magically became all the rest of life forms on earth seen throughout history!

https://www.youtube.com/watch?v=GjvuwneORrE&list=UUul300ELNnKXyxFosz1HyJA Now do you understand why I have written such notes as:

https://www.blastthetrumpet.org/PublicLetters/AAAUpdatedPublicAlertsMattersofLifeandDeath/Updates053016/Evolution%20and%20Atheism%20Intertwined%20Cults%20of%20the%20Insane.pdf - the cults of atheism and evolution have people in denial of reality. So much so they deny our Eternal Creator and as such are not representing themselves as having a sound mind.

https://www.youtube.com/watch?v=QbNpi2d5oEs

http://www.creationdino.blogspot.co.uk/2015/09/evolution-theorys-biggest-blow-wilsons.html?m=1

https://www.facebook.com/AnswersInGenesis/videos/10154140738429899/?pnref=story FOR THOSE WHO CLAIM THAT THESE THINGS ARE NOT BEING TAUGHT IN OUR PUBLIC SCHOOLS, OBSERVE THE TEXTBOOK QUOTATIONS IN THIS PRESENTATION:

https://www.youtube.com/watch?v=3HoHmSn5Gns and now do you understand why I want this nonsense removed from public education and that anyone practicing these cults of evolutionary thought and symbiotic atheism is demonstrating that they have become extremely

https://www.blastthetrumpet.org/PublicLetters/AAAUpdatedPublicAlertsMattersofLifeandDeath/Updates053016/Mentally%20Challenged.pdf; due to factual

https://www.blastthetrumpet.org/PublicLetters/AAAUpdatedPublicAlertsMattersofLifeandDeath/Updates053016/Brainwashing.pdf and therefore should NOT be allowed into teaching positions or media or other spheres of public influence? Consider the very existence of fossils; not just the fossils of large creatures, but of the small ones and of invertebrates, and even of cellulose fibers. Just the existence of these many fossils soundly disproves what they are teaching in the name of science regarding the fictionally created geologic column. https://youtu.be/so3BhFWaSeY Look at how fast decomposition of many species takes place:

https://www.youtube.com/results?search_query=dead+animals+decompose+rapidly

http://onlinelibrary.wiley.com/doi/10.1111/j.1469-8137.1941.tb07026.x/pdf

http://www.highveld.com/microbiology/biodegradation.html - scientists actually study what our Creator has made in order to gather enough knowledge to construct biodegradable materials themselves. The point of looking at how FAST decomposition takes place is to show that the very existence of fossils DOES NOT SUPPORT IN ANY WAY the deep time claims of the fictional "geologic column"; instead, they support RAPID death; by being covered by land and/or water AND SO MUCH LAND or WATER that other living organisms; observed ubiquitously, were not able to decompose the once living creature. (The very existence of mass millions of fossils buried and preserved in the global sedimentary layers CLEARLY SUPPORT the catastrophic Biblical Flood Account! they in NO WAY support (actually the fossils plainly FALSIFY) the nonsense of "millions of years" in the fictionally created geologic column; purporting claims of deep time layers of sediments in which those fossils are found).

https://www.google.com/search?q=fossils+of+fish+eating+fish&rlz=1C1GIGM_enUS535US535&tbm=isch&tbo=u&source=univ&sa=X&ved=0ahUKEwje1MPrpZDMAhWLxIMKHWrFAdEQsAQIHA&biw=1280&bi h=622 the vast majority of all fossils are marine fossils (over 95%), and in those are invertebrates and many creatures buried so fast that they were captured in the process of eating and giving birth. THINK about their claims! They pick up a rock and say things like this rock dates to be so many millions of years old, then they pick up another rock and claim that rock is hundreds of millions or even billions of years old, and yet another rock and claim it is only a few thousand years old. IF the elements in the earth all were hot like they teach, molten and then cooled etc. THEN THE MATTER THAT MAKES UP THE EARTH IS ALL THE SAME AGE; with the exception of asteroids space dust that may have fallen to earth at various periods of time. SO, THINK ABOUT THEIR CLAIMS THAT RADIOMETRIC DATING METHODS DATE THESE LAYERS TO BE MILLIONS OF YEARS APART! THEY ARE SO OBVIOUSLY LYING; I can't believe they are getting away with it!!!!!!!!

http://www.darwinthenandnow.com/2013/05/richard-dawkins-dumps-the-fossil-record/ - if darwin's theory was sound they're should be fossils detailing how evolution evolved from single celled to every other creature on earth BUT THERE ISN'T! Fully created multi-trillion celled creatures suddenly appear without any presumed evolution! Just as our Creator tells us in the Holy Bible!

https://www.youtube.com/watch?v=R1GJjHXX0Ic

The only reason people believe this nonsense is that they were told to believe it day in and day out all during their childhood, and then even the media reinforces the brainwashing with intentionally referring to anyone who would question this rubbish, as if we are the stupid ones! THIS EVIL BRAINWASHING MUST STOP!!!!!!!! It is incredibly EVIL to do to innocent little kids!!!!!!!! fossils inherently disprove deep time assumptions; that if it actually took that long for dead creatures to fossilize (while the layers of sediment millimeter by millimeter slowly rose up over their rotting corpses to bury them over the supposed "millions of years" it took to form those layers), they would have FAR SOONER completely decomposed; especially soft invertebrates like jellyfish

https://answersingenesis.org/kids/fossils/kids-feedback-how-do-jellyfish-fossilize/ and

https://www.google.com/webhp?sourceid=chromeinstant&rlz=1C1GIGM_enUS535US535&ion=1&espv=2&ie=UTF-8#q=how+do+jellyfish+become+fossils. Rather the scientific evidence proves fossilization can and does happen quite rapidly!

https://www.google.com/webhp?sourceid=chromeinstant&rlz=1C1GIGM_enUS535US535&ion=1&espv=2&ie=UTF8#

q=fossilization%20happens%20rapidly and particularly notice "when conditions are right".

https://www.google.com/search?q=artifacts+of+modern+petrification&rlz=1C1GIGM_enUS535US535&oq=artifacts+of+modern+petrification&aqs=chrome..69i57&sourceid=chrome&es_sm=93&ie=UTF-8 so then notice that those "right conditions" REQUIRES that the object be COVERED by other liquids or solids; to provide the elements that replace the original organic structure and cause the metamorphosis! (Metamorphosis BEFORE decomposition PROVES that fossils DO NOT, ABSOLUTELY DO NOT, take "millions of years" to form; but rather MUST form RAPIDLY!

https://www.youtube.com/watch?v=yH35OQbV7GY and

https://www.youtube.com/watch?v=d99iLtzZKZ0)
Therefore, the very existence of ALL fossils, FALSIFY what is being taught regarding deep time; slow sedimentary ages of layers in the earth's crust! HUGE (polystrate) fossils and petrified trees that span multiple layers of the fictional "geologic column" WOULD HAVE DECOMPOSED, OXIDIZED (if they were not covered)

https://www.youtube.com/watch?v=k3VOum6x91w, AND OTHERWISE ERODED TO DUST AND MINERALS LONG BEFORE the "millions and millions of years" evolutionists would have you believe are represented by those layers; LONG BEFORE! -

https://www.google.com/search?q=petrified+trees+vertical+through+many+layers&rlz=1C1GIGM_enUS535US535&es_sm=93&biw=1066&bih=702&tbm=isch&tbo=u&source=univ&sa=X&ei=BC5GVLXL4y2yASj64C4Bg&ved=0CB0QsAQ and

https://www.google.com/search?q=Polystrate+fossil&rlz=1C1GIGM_enUS535US535&oq=Polystrate+fossil&aqs=chrome..69i57j69i60&sourceid=chrome&es_sm=93&ie=UTF-8#q=Polystrate%20fossils) Which is more REASONABLE, that MIRACULOUSLY ALL THOSE FOSSILS, INDIVIDUALLY DIED, AND SUBSEQUENTLY WAS MYSTERIOUSLY (without rational explanations for all the millions of fossils found to date) BURIED BY LAND AND/OR WATER SO FAST AS TO PREVENT ORGANIC DECOMPOSITON, AND THAT THOSE MIRACULOUS EVENTS OCCURED MILLIONS OF TIMES, OVER MILLIONS OF YEARS, FOR EACH OF THOSE INDIVIDUAL FOSSILS OR THAT THE SEDIMENTARY LAYERS FOUND WORLDWIDE; IN WHICH ALL THESE DEAD CREATURES, EVEN SOFT INVERTEBRATES, WERE CATASTROPHICALLY COVERED IN A GLOBAL FLOOD

WHICH HYDROLOGICALLY SORTED THE SMALLER AND LARGER SEDIMENTS, AND CREATURES ALL AT ONCE?!

https://www.youtube.com/watch?v= QwOjXvD5DQ (Keep in mind RECORDED HISTORY has the flood account in hundreds to thousands of cultures from all over the world -

https://www.youtube.com/watch?v=lktmmd7YnD8;

Just use a search engine if you want to read up on all those records). My POINT HERE, is that evolutionists and atheists CLAIM their ideologies are scientifically supported AND reasonable and that the Creationists are dumb and ignorant; who blindly believe their religions; exclusively by faith; with no rationale, reason or scientific evidence; when ANYONE CAN PLAINLY SEE the converse is true. Creationists value HISTORY, more importantly the provably and verifiably DIVINELY INSPIRED HISTORY in the Holy Bible! That history that the One True God, Jesus Christ confirmed (and still confirms) is much more important than any fabricated opinions from anyone over imaginary billions of years ago. The observable EVIDENCE of the earth's crust comprised of the same sedimentary layers globally; in which mass millions of species were suddenly buried supports rapid hydrologic sorting; NOT creatures dying millions and millions of years apart and slowly fossilizing, while sediments are slowly formed all around and over them; despite everything we observe regarding the organic cycle on earth.

https://www.google.com/search?q=organic+cycle+on+earth%2C+life%2C+death%2C+decomposition%2C+life&rlz=1C1GIGM_enUS535US535&oq=organic+cycle+on+earth%2C+life%2C+death%2C+decomposition%2C+life&aqs=chrome..69i57&sourceid=chrome&es_sm=93&ie=UTF-8

https://www.youtube.com/watch?v=jvaNP5xH4HI

http://www.creationscience.com/onlinebook/Liquefaction7.html - Creationists are the rational and reasonable people who base their worldview on EVIDENCE and HISTORY as opposed to irrational evolutionists who are entirely faith based in imaginary nonsense that no one has ever seen or heard - billions of years ago! NONSENSE! It is the evil-u-shunist that are promoting their cultic beliefs on innocent children in the name of science; NOT THE BIBLICAL CREATIONISTS! The BIBLICAL CREATIONISTS are just teaching the TRUTH!

https://www.youtube.com/watch?v=hmMIpgIyOnY

https://www.youtube.com/watch?v=_Qw0jXvD5DQ

https://www.youtube.com/watch?v=yH35OQbV7GY

http://www.creationevidence.org/carlbaugh/v2ch1.htm - experiments have been done proving the hydrologic sorting that took place on the flood.

https://www.google.com/search?q=carl+baugh+fast+fossils+pickles+and+cowboy+boot&rlz=1C1GIGM_enUS535US535&es_sm=93&biw=1066&bih=702&source=lnms&tbm=isch&sa=X&ei=8WkVZynIcv7sAWVl4DwBQ&ved=0CAcQ_AUoAg

http://regentsprep.org/regents/core/questions/questions.cfm?Course=ESCI&TopicCode=04the global flood

https://www.youtube.com/watch?v=lktmmd7YnD8 more rationally accounts for the geological deposits and fossil layers we observe globally. As such, fine sediments (smaller size creatures would have been trapped in the finer silt deposits; while larger creatures settled above them) https://www.youtube.com/watch?v=hmMIpgIyOnY Since the Bible and prayer were forcibly removed from public education by corrupt judges and those that fund public education; which has had a devastating effect on Americans and our nation as a result - http://www.whatyouknowmightnotbeso.com/graphs.html

http://www.blastthetrumpet.org/PublicLetters/AAAUpdatedPublicAlertsMattersofLifeandDeath/MoreUpdates031315/THE%20PLANNED%20DEMORALIZATION%20OF%20AMERICA.pdf We went to school together for the most part. (Not to boast here, but I have lost all patience in dealing with arrogant atheists and evolutionists who sneer at Creationists with unjustified slurs and assumptions about our intelligence.) I was the kid in school that never studied because I never needed to, that pointed out errors in the text books, that got top grades without trying, that read the text books in the first few days of each academic year; during many redundant lectures, that completed perfectly the timed entrance exam for military recruitment; something those recruiters had never encountered, that was nominated to both West Point and Annapolis Military Academies, SEALs/EOD entrance qualified, finished a six month self paced course in two weeks with a 98% average; earning a meritorious commendation from the base commander of the Great Lakes Naval Training Center, years of diversified studies in colleges with no needed study through it all and over a 3.9 GPA, ordination and licensing as a minister of the Gospel of Yahoshuah Am Meschiach, aka Jesus the Christ, I read and study virtually every waking moment that I am healthy enough to do so and I can tell you that when I encounter evolutionists and atheists online, they are so thoroughly indoctrinated that; even while arguing in support of one of the many versions of the "big bang" theory, they will attempt to insult me with logic fallacies simply for rephrasing what they claim to believe; that the universe has an origin therefore (time, energy and matter is expanding in space from that origin -

https://www.facebook.com/photo.php?fbid=451991028213195&set=a.115635768515391.22520.100002069048072&type=3&theater; just read some of the comments against me for only stating what I saw about how the matter in the universe came to be:

https://www.youtube.com/watch?v=ri3R6vFVfEg&feature - if you read the comments below the video, you will see that in the dream in which I saw the possible beginning of the new heavens and earth, that because it so aptly fits what we observe in our universe, I concluded that it was also how He formed the present heavens and earth; but that might not be so.

https://www.blastthetrumpet.org/PublicLetters/AAAUpdatedPublicAlertsMattersofLifeandDeath/Updates053016/Th

e%20Space%20Time%20Continuum.pdf HOWEVER, the Book of Enoch seems to confirm what I saw. http://www.sacred-texts.com/bib/boe/); (so many versions of the "Big Bang" theory it's not surprising that none of them understand the suppositions and connotations of what they champion as scientific knowledge -

https://www.blastthetrumpet.org/PublicLetters/AAAUpdatedPublicAlertsMattersofLifeandDeath/Updates053016/So%20You%20Think%20the%20Big%20Bang%20is%20Scientific.pdf) What "evolutionists" do is they look at bones they dig up and find a smaller hominid test its age, look at skull shape, etc. and then put a fancy Latin name on it and CLAIM it is a transitional form of our supposed human evolution; when you can find LIVING examples on planet earth of all their so called varied speciations along their atavistic imaginations of fictional evolution:

http://jesus-is-savior.com/Evolution%20Hoax/Evolution/12a.htm - they're diggin' up bones and tellin' stories

https://www.youtube.com/watch?v=F6M2YuAVxuQ - thinking they're fictional imaginations, fanciful stories, are "scientific facts" is absolutely ridiculous; extremely foolish.

https://www.google.com/search?q=living+pygmy+tribes&rlz=1C1GIGM enUS535US535&espv=2&biw=1066&bih=702&tbm=isch&tbo=u&source=univ&sa=X&ei=fwKKVMv8DsSdygSuu4LwDw&ved=0CB0QsAQ &dpr=1.5 - humans come in all kinds of shapes and sizes.

https://www.google.com/search?q=samoan+giants&rlz=1C1GIGM enUS535US535&espv=2&biw=1066&bih=702&source=lnms&tbm=isch&sa=X&ei=dQOKVJXNMIS9yQTzioCwCA&ved=0CAcQ AUoAg

https://www.google.com/search?q=samoan+giants&rlz=1C1GIGM enUS535US535&espv=2&biw=1066&bih=702&source=lnms&tbm=isch&sa=X&ei=dQOKVJXNMIS9yQTzioCwCA&ved=0CAcQ AUoAg#tbm=isch&q=living+giants - many things in history and to this day DEFY evolutionary constructs.

https://www.google.com/search?q=living+dwarfs&rlz=1C1GIGM enUS535US535&es sm=93&source=lnms&tbm=isch&sa=X&ei=KgSKVIuNGY2ryATGnYD4Aw&ved=0CAgQ AUoAQ&biw=1066&bih=702

https://www.google.com/search?q=living+people+smaller+than+dwarfs&rlz=1C1GIGM enUS535US535

&oq=living+people+smaller+than+dwarfs&aqs=chrome..69i57&sourceid=chrome&es_sm=93&ie=UTF-8

So INSTEAD of "evolution", what we OBSERVE is widespread genetic variations within species that some "scientists" would CLAIM are "transitory" forms along their fictional evolutionary constructs; IF such LIVING variations were the bones they were digging up. WHAT WE OBSERVE shows information over time is LOST; NOT gained! (in other words, "mutations" are a bad thing; not beneficial!) such extreme mutations DO NOT survive but inevitably die off! So EXTINCT genetic abnormalities among ANY and ALL species; including hominids, DO NOT show "evolution" but rather mutations that have become extinct over time as any reasonable scientist expects based on the EVIDENCE!

https://www.facebook.com/WHYOutreach/photos/a.651187234910638.1073741829.210188172343882/933666296662729/?type=1&theater

https://www.youtube.com/watch?v=GjvuwneORrE

There is a need here to interject that the current "falsification" process of scientists submitting their theories only to respective scientific peers and disciplines is inherently flawed and it has been so throughout history.

Peer review publications are controlled by a few persons with obvious agendas; so those who present dissenting opinions based on observable evidence; like myself are handily dismissed and even laughed at by those who have been successfully demoralized to a state where they are incapable of critical thought any longer. So, you actually find people believing against all logic and reason; the very foundational laws of science and knowledge, ridiculous notions like "everything came from nothing". Conformity has been so ingrained into these persons that they simply find it unthinkable to challenge the status quo; and those who do often find their careers decimated. Ask yourselves WHY? I will TELL you why. It is because WHOEVER IS IN POWER, WHOEVER CONTROLS THE MONEY AND THE MILITANT MIGHT, rules! (at least they exercise such undue influence that people are AFRAID to dissent) For example,

https://www.blastthetrumpet.org/PublicLetters/AAAUpdatedPublicAlertsMattersofLifeandDeath/Updates053016/Bad%20Cops.pdf - Thank God for all the men in uniform doing their best to live Godly, Righteous Lives!

But when they go bad proves that the law doesn't matter in the moment a cop decides because they have a gun and a badge, they can do whatever they want without consequence; including murdering innocent citizens. And our entire government and governments all over the world and down through history all prove that whoever is in power and authority will enforce their own wills upon the masses forcefully; with RARE exceptions of those who realized GOD ALMIGHTY ultimately governs us all and that they will be held responsible for their actions sooner or later. In the scientific community, it is seen in the way they all defer to whatever person or persons are paying them! (In other words, a "consensus" among the scientific community, PROVES NOTHING, but that the MONEY behind their current popular opinions (SO-CALLED "theories" and "laws"); is FORCING that consensus. Science in such circumstances is no longer science and the "falsification" process FAILS! (because those who would dissent or offer critical evidence against the popular and accepted theories or laws, FEAR for their careers and hence their personal survival and that of their families; like all the dissenting creationists that have been fired in recent history). As such, peer to peer review submission becomes virtually MEANINGLESS. The BEST way to know whether or not ANYTHING is fundamentally true is to KNOW the ONE who is TRUTH.

https://www.blastthetrumpet.org/PublicLetters/AAAUpdatedPublicAlertsMattersofLifeandDeath/Updates053016/HOW%20TO%20KNOW%20TRUTH%20IN%20A%20WORLD%20OF%20LIES.pdf - Know Jesus Christ to Know TRUTH!

But if you REALLY want the falsification process to work, you absolutely CANNOT fire other scientists who express a critical opinion of a theory or law; NO MATTER WHO submitted that theory or law; NO MATTER how famous they are; NO MATTER how many degrees they have! If you are a TRUE SCIENTIST, then EVIDENCE submitted to you should not matter WHO presents it! An apple is an apple whether or not a post graduate hands it to you or an infant, whether or not a genius presents it to you or an imbecile! So, when I say LOOK at the EVIDENCE, I mean it! Otherwise flawed "theories" and "laws" remain in place and HINDER the progress of science and scientific discoveries as they are forcefully ingrained upon generation after generation as "scientific fact" and valid; regardless of whether or not they actually are. PLEASE READ MY NOTES AND LOOK AT MY CITATIONS IN THEM! WATCH THE DOCUMENTARIES, READ EACH SO CALLED "MEME" BECAUSE THOSE youtube, google, and facebook citations are linked to FACTS (the memes have citations in the paragraphs describing them; linked to history and scientific discoveries, the google citations have pages and pages of scientific journals, historical research, expert opinions, verifiable observations, etc. and youtube has documentaries with biographies cited within or in the text below the videos; or are otherwise verifiable in the worldwide database of the Internet). I DO CITE the Holy Bible as scientific evidence for many reasons,

https://www.blastthetrumpet.org/PublicLetters/AAAUpdatedPublicAlertsMattersofLifeandDeath/Updates053016/Th

e%20Ineffably%20Awesome%20Holy%20Bible.pdf; The Holy Bible is ineffably AWESOME because it comes from our Eternal Creator!

but namely to prove the contents are factually supported by what we OBSERVE; by SCIENCE! I also DO cite the findings of scientists, and even persons who are not in the OFFICIAL capacity of scientists, simply because NO ONE has to be an OFFICIAL scientist to PERCEIVE REALITY; to ACCURATELY OBSERVE the universe and all therein. (EVERYONE who ACCURATELY observes creation is a "scientist" whether or not they are indoctrinated or degreed as such!) In fact, my research shows the general population may be more "scientific" today, than many degreed professionals for the aforementioned reasons and the very fact that public education has provably become public mass brainwashing. So, possession of multiple degrees tells the rest of us that such persons have been even MORE brainwashed into conformity with popular "scientific" notions, than those without them! ALL scientists should UNITE when fellow scientists EVEN ONE fellow scientist is fired simply for presenting a dissenting opinion or critique of any theory or law; no matter how popular it is; otherwise, "falsification" fails, and "science" is thereby turned into nothing more than popular fiction! (Issuing a dissenting opinion DOES NOT entitle teachers, scientists or media professionals to KNOWINGLY DECEIVE AND LIE TO STUDENTS! - such as the thousands of deceptions that have been proven fraudulent in educational text books but still remain in print and are still being forced upon the unsuspecting masses and their children)

https://www.youtube.com/watch?v=b8GgrUposII

Because of this fact (successful demoralization), I have often met those who REFUSE to read my notes and look at the evidence say ridiculous things like, "so you think you're smarter than the entire scientific community over the past few centuries!" and I think NOTHING of the kind; only that as I have STATED, EVIDENCE SHOULD NEVER BE IGNORED, NO MATTER WHO PRESENTS IT; and I am not the only critic of modern theoretical physics and evolutionary fiction as taught the public all over the world! IT IS BECAUSE, I have no fear of anyone or anything but GOD ALMIGHTY, that I can boldly make a fool of myself (in the evolutionist's thinking and secular mindset); because I KNOW we are ALL students before our Lord Jesus Christ and we ALL know very, very little! It is also because I think it is VERY EVIL to sneer in derision at those who hold to Biblical Truths that have been declared for millenniums and stood the test of time and the greatest minds in the history of the world; only because those doing so THINK science and reason is in opposition to; rather than, PLAINLY SUPPORTS what the Word of our Eternal Creator; as recorded in the Holy Bible clearly tells us all. In other words, I SPEAK BOLDLY; not out of arrogance, but out of REAL and GENUINE concern for the lives of all mankind and all life on the planet; that it is WRONG to deceive people young or old into thinking something is scientific when it is NOT and VERY WRONG and EVIL to keep people; in the name of science and reason, from KNOWING WHAT IS ACTUALLY SCIENTIFIC AND REASONABLE; to knowingly deceive people in the name of science and reason, to keep them from KNOWING TRUTH; to brainwash them into believing lies and fiction in the name of science; in order to keep them from KNOWING GOD ALMIGHTY our CREATOR and LORD of the Universe!!!!!!!! I have read many of the latest scientific discoveries and watched more documentaries than I can enumerate, findings and submissions over the years in many disciplines, but in decades of research I have not yet found any scientific justification for the likes of the arrogant derision persons like Richard Dawkins and Neil deGrasse Tyson espouse toward Biblical Creationists; while making cosmologically and evolutionary insupportable claims that roll so smoothly off their tongues.

https://www.blastthetrumpet.org/PublicLetters/AAAUpdatedPublicAlertsMattersofLifeandDeath/Updates053016/The%20Dawkins%20Delusion.pdf It is physically nauseating to listen to their monumental ignorance; while they are obviously in such narcissistic affinity for what they THINK they know! I HAVE read the deep time claims and the vast distance claims, BUT I contend that EVERY SINGLE stellar dating method (similar to the extremely inaccurate and fallacious radiometric dating methods) I have researched have whole lists of ASSUMPTIONS in them of varying degrees (can be falsified) with the exception of the closer celestial bodies that the parallax method can measure. (Dating methods that go beyond verifiable recorded history all have greater and greater amounts of error in them the further deep time claims are asserted).

http://www.cs.unc.edu/~plaisted/ce/dating.html and

http://creation.com/the-way-it-reallyis-little-known-facts-about-radiometric-dating and

https://answersingenesis.org/geology/radiometricdating/radiometric-dating-problems-with-the-assumptions/ and

http://biblicalgeology.net/blog/fatalflaw-radioactive-dating/ and

https://www.google.com/webhp?sourceid=chromeinstant&rlz=1C1GIGM_enUS535US535&ion=1&espv=2&ie=UTF8#q=errors%20and%20assumptions%20in%20stellar%20deep%20space%20deep%20time%20dating%20 methods (scoffers were prophesied to come in the end times and make false assertions in the name of science -

http://biblehub.com/2_peter/3-3.htm and
http://biblehub.com/2_peter/3-5.htm and
http://biblehub.com/jude/1-18.htm and
http://biblehub.com/1_timothy/6-20.htm -

Their "science" is actually deceptive and fraudulent and is what the entire fiction of the modern evolutionary paradigm is built upon) ACCURATE OBSERVATIONS, REAL SCIENCE, leads us all to TRUTH, leads us all to our CREATOR! "The most appropriate method to obtain accurate stellar ages relies on the computation of stellar models. However these models are far from being perfect, they are affected by several sources of uncertainties." - http://arxiv.org/pdf/1410.5336.pdf

Do you comprehend that ALL CURRENT MODELS are based on the fictional evolutionary constructs?! Do you really UNDERSTAND that from the very onset of atheistic and evolutionary cosmology; evolutionary constructs of how the chemicals formed, and the stars is ALL BUILT ON FALSE ASSUMPTIONS?! As such, BOTH with deep time dating methods on earth and in astrophysics, I find the above statement to be true; if you can actually get the scientists to ADMIT it! key phrase "...FAR FROM BEING PERFECT, they are affected by SEVERAL SOURCES OF UNCERTAINTIES." and YET it is astonishing how many people walk around laughing at Biblical Creationists saying don't you know? it's a "scientific fact!" and "dating methods are highly accurate!" BASICALLY, WHEN SCIENTISTS ARE HONEST, these dating methods ARE NOT inerrant, ARE NOT accurate, ARE NOT infallible, ARE NOT perfect, ARE NOT reliable, they are simply the best they have at the moment and are at best only attempts to approximate these things; and BECAUSE OF WHO IS FUNDING them tends to support the ideologies of whoever is ensuring their own job security and survival at the moment; and at worst WILDLY inaccurate; so much so, as to be practically useless and even DECEPTIVE! When anyone looks into these things currently and historically, it is plain to see that, CONTRARY TO POPULAR OPINION, "SCIENTISTS" ARE NOT ALWAYS ABLE TO BE AS OBJECTIVE as the masses might think, and even when they are free to be such, all people I have encountered and read about in history are VERY subjective; even and especially those claiming objectivity.

http://www.creationconcepts.org/resources/DATING-2.pdf The further one trespasses beyond recorded history, the more assumptions come into play and with each assumption, the greater chance of error increases! I AM NOT CONTENDING FOR A YOUNG UNIVERSE; I AM ONLY SAYING THAT THOSE WHO THINK DEEP TIME IS A "SCIENTIFIC FACT" AND ALL THESE DATING METHODS ARE INERRANT HAVE NOT READ THE SCIENTIFIC FINDINGS! THEY ARE REPEATING THEIR INSTITUTIONALIZED

BRAINWASHING! AND I HAVE HAD IT WITH THESE PEOPLE WHO KEEP CLAIMING SUCH NONSENSE IS "SCIENTIFIC FACT" ONLY BECAUSE THEY HEARD IT FROM A PROFESSOR OR READ IT IN A TEXT BOOK OR SAW IT ON TV! The dissenting creationists HAVE NOT been heard and are increasingly being given no opportunity to point out all these many things concerning the popular paradigms of evolutionary thought are FAR MORE fiction and assumptions, speculations, and imaginations than ANYTHING that truly meets the merits of science, or "scientific fact", "theories", "laws" or postulates or even a well thought out hypothesis! So, the assumption that those supporting Intelligent Design have not heard or learned what evolutionary theories teach is false. I challenge therefore evolutionists to give Biblical Creationists NOT decades of your lives, like we were forced to, in listening to such wild imaginations; but to take a very small time, proportionately; considering thoughtfully the evidence that is being censored from public education by viewing some of their presentations and documentaries.

https://www.youtube.com/watch?v=Gjvuwne0RrE&list=UUul300ELNnKXyxFosz1HyJA

https://www.youtube.com/watch?v=8xL6sibLb6o&list=PL6Q7oDH9-hiiBa4RmUJoh-jojfJT13r8Q - evolutionists immediately began to make excuses for the discovery of human footprints with dinosaur tracks. They claimed that they aren't human footprints, but just eroded dinosaur tracks. Then they claimed someone carved the tracks in the stone by hand. (was so expert that they did it in a matter of days complete with proper anatomical features and weight distribution evidence) In fact, there were so many who challenged the empirical evidence of the find, that Dr. Baugh went back, while the cameras were rolling live, and continued to remove the ground cover and uncovered MORE human footprints and dinosaur tracks in the SAME LAYER!

https://www.youtube.com/watch?v=q8rh2Zp6wc8 - notice he acknowledges the difference between the fake, carved prints and the genuine.

http://www.talkorigins.org/faqs/paluxy/onheel.html

https://www.youtube.com/watch?v=HaQqa90e9KQ - seeing is believing. (evolution as it is taught today is FALSE!) -

https://www.youtube.com/watch?v=7xYyHU5mRFo

https://www.youtube.com/watch?v=dWe3cteDuBc

https://www.youtube.com/watch?v=bZzIihVUti4 - the "geologic column" as is forced on students by the mass millions, is FACTUALLY fiction! -

https://www.google.com/webhp?sourceid=chromeinstant&rlz=1C1GIGM_enUS535US535&ion=1&espv=2&ie=UTF8#q=charles+lyell+created+the+fictional+geologic+column&tbm=vid and

https://www.google.com/webhp?sourceid=chromeinstant&rlz=1C1GIGM_enUS535US535&ion=1&espv=2&ie=UTF8#q=the+geologic+column+is+found+nowhere+on+earth&spell=1

If it was JUST the hotly debated footprints that invalidate evolutionary constructs, then we might argue indefinitely over them; but there is FAR MORE EVIDENCE that dinosaurs and man coexisted.

https://www.youtube.com/results?search_query=dinosaurs+and+man+coexisted - plenty of evidence that we coexisted with dinosaurs.

https://www.google.com/search?q=dinosaurs+and+man+coexisted&rlz=1C1GIGM_enUS535US535&oq=dinosaurs+and+man+coexisted&aqs=chrome..69i57j69i60&sourceid=chrome&es_sm=93&ie=UTF-8 - loads and loads of evidence.

https://www.google.com/webhp?sourceid=chromeinstant&rlz=1C1GIGM_enUS535US535&ion=1&espv=2&ie=UTF-8#q=living%20dinosaurs

https://www.google.com/webhp?sourceid=chromeinstant&rlz=1C1GIGM_enUS535US535&ion=1&espv=2&ie=UTF8#q=mammal%20and%20dinosaur%20fossils%20together - buried altogether.

https://www.google.com/webhp?sourceid=chromeinstant&rlz=1C1GIGM_enUS535US535&ion=1&espv=2&ie=UTF8#q=list%20of%20soft%20tissue%20dinosaurs%20found - soft tissue dinosaur specimen date to recent history, not millions of years ago.

http://www.biblebelievers.org.au/giants.htm

https://www.facebook.com/icr.org/photos/a.1015026820
8564451.345328.43552419450/10152879620
299451/?type=1&theater

https://www.youtube.com/watch?v=_JV6K1az6NM&list=P
L6Q7oDH9-hihPJTH_wAWIk-rduZzD1d-f

https://www.youtube.com/watch?v=vSwt8edtHUA&list=P
L6Q7oDH9-hihPJTH_wAWIk-rduZzD1df&index=10

https://www.youtube.com/watch?v=ezG1lPFUMI4

https://www.youtube.com/watch?v=EWp0ewX3W94#t=3
703 - artifacts are found proving humans and dinosaurs coexisted, and yet STILL instead of questioning dating methods (which are provably false) people INSTEAD jump to the conclusion that humans have been around for "millions" of years. The BULK of the EVIDENCE proves the GLOBAL FLOOD and RECENT history! (Dinosaurs found with skin and soft tissue even digestive remains! -

http://www.dcclothesline.com/2014/03/08/contradictingevolution-massive-dinosaur-soft-tissue-discovery-china-includes-skin-feathers/ (this was proven to be a scam - ALL OF THEIR SO-CALLED missing links HAVE BEEN PROVEN TO BE SUCH! evolution is impossible! It's time people recognized it so they STOP trying to brainwash little kids with that rubbish!)

https://www.google.com/webhp?sourceid=chrome-instant&rlz=1C1GIGM_enUS535US535&ion=1&espv=2&ie=UTF8#q=chinese%20scams%20the%20smithsonian%20with%20false%20fossil%20and%20bird%20feathers) but modern birds have been fossilized in the same layers as dinosaurs that evolutionists would have you believe predated them by "millions of years" -

https://www.google.com/webhp?sourceid=chromeinstant&rlz=1C1GIGM_enUS535US535&ion=1&espv=2&ie=UTF8#q=modern%20bird%20fossils%20found%20with%20dinosaurs and

https://www.google.com/search?q=fossilized+dinosaur+found+nearly+complete+with+skin+and+digestive+remains&rlz=1C1GIGM_enUS535US535&es_sm=93&biw=1066&bih=702&source=lnms&sa=X&e

i=v5tGVKqzIuTfsASbpIDwBg&ved=0CAUQ_AUoAA&dpr=1.5#q=hadrosaur+mummy&tbm=vid and

https://www.youtube.com/watch?v=7Vvr46hz0hk)

https://www.youtube.com/watch?v=R6GiNQvugn0

https://www.youtube.com/watch?v=8DXs3B9uEuI&list=PL126FC84AA7438000

https://www.youtube.com/watch?v=lktmmd7YnD8&list=PLEA099CE59F0A0322

https://www.google.com/webhp?sourceid=chromeinstant&rlz=1C1GIGM_enUS535US535&ion=1&espv=2&ie=UTF-8#q=megalania%20sightings

https://www.google.com/search?q=mokele+mbembe&rlz=1C1GIGM_enUS535US535&oq=mokele+mbembe&aqs=chrome..69i57j69i61&sourceid=chrome&es_sm=93&ie=UTF-8

https://www.youtube.com/results?search_query=lake+monsters+worldwide

https://www.youtube.com/watch?v=_dMi_xqnMas

https://www.youtube.com/watch?v=P-B2hACS0dQ - whether cosmic evolution (they have no REASONABLE EXPLANATION for the existence of the universe and NO OBSERVABLE, REPEATABLE EXPERIMENTS to support the many forms of their so called "scientific theories" regarding evolutionary cosmology, chemical evolution (they have no REASONABLE EXPLANATION for the formation of the elements we have discovered to date, let alone ALL the elements that may yet be discovered in the unexplored universe), stellar evolution (despite CLAIMS otherwise NO ONE has SEEN the formation of stars, planets, galaxies EXCEPT our Creator and those souls He may have shown such things to. Again, no scientific repeatable experiments to demonstrate just how stars formed; but even if there were observations of merging smaller masses into larger masses, this wouldn't be a "birth of a star" or galaxy but simply the kind of fluid dynamics we observe on the space station happening on a larger scale); organic evolution (fictional and unscientific; inherently known as abiogenesis - this is AT THE ONSET positively ludicrous! The Law of Biogenesis states plainly WHAT HAS BEEN OBSERVED "Life comes from life!" Again, science; WHAT WE OBSERVE, supports the Biblical Creationists; NOT the evolutionists), and the Theory of Evolution, as taught mass millions in the name of science is SCIENTIFICALLY INSUPPORTABLE! (Imaginations, FICTION - THINK ABOUT WHAT THEY ARE TEACHING AND HOW IT DEFIES SCIENTIFIC LAWS THROUGHOUT! (Their entire constructs are provably fictional - SCIENCE, WHAT WE OBSERVE, disproves/falsifies virtually in every way what they are teaching! -

https://www.youtube.com/results?search_query=evolution+is+stupid,

https://www.youtube.com/results?search_query=evolution+lies+in+textbooks),

https://www.facebook.com/photo/?fbid=6879704782108422&set=a.115635768515391 - when mammals attacked dinosaurs.

The premises are false, the arguments are false, the entire evolutionary constructs and ridiculous conclusions have been THOROUGHLY FALSIFIED by actual science! On the other hand, science DOES SUPPORT THE BIBLICAL ACCOUNT THROUGHOUT! This is a most SERIOUS issue! matters of life and death and everlasting destinies are at stake! People are getting fired merely for questioning Darwinian evolution! (Scientists CLAIM they present their theories and laws publicly to WELCOME criticism; so that knowledge is increased; but in REALITY, people are losing their livelihoods! careers! ability to survive in our modern society; who do just that.) The state, in clear and overt violation of our highest laws, is provably supporting a religious worldview; dogmatically and forcibly coercing the fictional evolutionary constructs against all the scientific support for Intelligent Design).

https://www.blastthetrumpet.org/PublicLetters/AAAUpdatedPublicAlertsMattersofLifeandDeath/Updates053016/The%20cult%20of%20modern%20scientism%20versus%20the%20Science%20of%20Biblical%20Creation.pdf - darwinism is not science! pure fantasy!

These people make statements like "but we make predictions" and then claim THOSE DESIGNED "experiments" and "methods" prove their predictions and offer those intentional frauds as "proof" for their theories or so-called scientific laws. (Like how many STRONGLY assert all the radiometric dating methods as being highly accurate; when THAT IS NOT A TRUE STATEMENT! -

https://www.youtube.com/watch?v=TVuVYnHRuig

and ALL of those kinds of dating methods make NUMEROUS ASSUMPTIONS; that is NOT "scientific"! -

https://www.blastthetrumpet.org/PublicLetters/AAAUpdatedPublicAlertsMattersofLifeandDeath/Updates053016/Bill%20Nyes%20Science%20Lies.pdf - Just ask yourselves when ANYONE claims deep time; half-lives or otherwise, HOW LONG did they factually sample the radioactive substance? HOW many samples or permutations did they run to validate their findings? What were the conditions of the tests run on the samples (natural or unnatural settings, environment), what manmade substances were used in the procedure? and keep in mind that the idea of decay of these complex elements is yet another observation that FALSIFIES chemical evolutionary fictional constructs of how less complicated elements somehow formed themselves into all the more complex elements; BECAUSE those elements are DECAYING; and thus, EVERYTHING WE CAN ACTUALLY OBSERVE falsifies the fictional constructs of evolutionary theory throughout!) I was in a debate recently online in which a person actually claimed that Stephen Hawking, in his opinion was the "smartest man on the planet" and then quoted Hawking, " Because there is a law such as gravity, the universe can and will create itself from nothing..." - http://en.wikipedia.org/wiki/The_Grand_Design_(book) All sane, rational, reasonable persons left on earth KNOW that statement is FALSE (not scientific) and not even science fiction; but is purely imaginary; on a level that can be called delusionary; if the statement is seriously given in the name of science! In fact, it is so completely false that it is probable no actual "scientist" or rational person would make such statement; so those famous scientists that have, are more than likely conducting social experiments to see how much influence these persons (scientists; especially famous ones) carry upon the public in general. I would hope such intelligent persons would NOT participate in such experiments, as history proves that such mass experiments are typically funded by such evil persons who intend to use the findings for manipulating the masses with planned social engineering (mass brainwashing).

http://www.livescience.com/28132-what-is-nothing-physicists-debate.html - when denying the clear definition of a word ("nothing" is the absence of "everything", the very opposite, the antonym, it is MOST DEFINITELY not synonymous or the cause of ANYTHING; by definition!) and its connotation globally; because you are attempting to do the impossible (declare Creation exists without the Cause of our Eternal Creator,

https://www.blastthetrumpet.org/PublicLetters/AAAUpdatedPublicAlertsMattersofLifeandDeath/Updates053016/GOD%20Part%201.pdf and https://www.blastthetrumpet.org/PublicLetters/AAAUpdatedPublicAlertsMattersofLifeandDeath/Updates053016/G

OD%20Part%202.pdf); then IT IS CLEARLY APPARENT TO ALL; that these persons are NOT practicing the scientific method and are engaging in imagination, fantasy, fiction NOT science! Everyone but the most deceived and self-deluded KNOWS that something cannot come from nothing BY DEFINITION! and by the Laws of Epistemology that are INSTANTLY VIOLATED by such a contradictory claim! (If such persons are SERIOUSLY asserting such nonsense, then about the only actual scientific evidence they are presenting to the world is that they have gone mad/insane; totally and completely! - and further support the Word of GOD which has stated for millenniums - http://biblehub.com/romans/1-28.htm - "And just as they did not see fit to acknowledge God any longer, God gave them over to a depraved mind..." and their arrogance simultaneously presented to the world with such massive ignorance is further fulfillment of http://biblehub.com/romans/1-22.htm - "Professing to be wise, they became fools,..." This notion of everything coming from nothing is not only perhaps one of the most ludicrous statements ever to be made by anyone, let alone in the name of science; so much so, that it flies past laughable immediately to being extremely pitiable. (BUT UNDERSTAND THAT THIS IS THE END RESULT OF ANYONE WHO EMBRACES THE MODERN evolutionary paradigm; in part or in whole.)

http://plato.stanford.edu/entries/aristotle-noncontradiction/

Far too many modern "scientists" redefine words, postulate ridiculous notions that are not supported by observation, and do such things as intentionally design tests, falsify test results, or ignore facts, because of peer pressure, job security, (the funding forcing them to say such things; as they know are NOT scientific) and for other unacceptable reasons that are provably damaging the millions of innocent people who go on thinking they are educated and KNOW something as true science that is actually nothing but fiction and wild imaginations or fantasies. They do all these deceptive things in vain attempts to support such foolish notions; WHILE THEY CLAIM THE VERY AXIOM OF SCIENCE demands that they should instead be looking for ways to falsify those theories or laws and finding none, THEN conclude that they have reasonably supported their "theory", hypotheses, or "laws". BUT THEY DO NOT DO THIS BECAUSE ANYONE CAN MAKE ANYTHING LOOK FALSE OR DECEIVE THE PUBLIC; as magicians prove to the whole world with all their illusions. So, throw out their FALSE "axiom of science" and the whole notion that falsification plays any part whatsoever in determining what is actual science, actual knowledge and actually true. TRUE observations and FACTS can be presented in such a way or interpreted in such a way as to SEEM to falsify something; as is widely done, willfully, by those who misinterpret the Holy Bible in a vain attempt to claim something in it is in error. (Like I said clever people can "falsify" or rather appear to falsify ANYTHING simply through deceit and since we presently live in a world of open mass deceptions, liars, deceivers, ignorance, etc. that whole method is a very poor suggestion to prove a scientific proposal of any kind.) INSTEAD, people MUST ATTEMPT to be HONEST, ACCURATE, OBSERVERS of REALITY, and ACKNOWLEDGE the EXISTENCE of REALITY, even if it is unpleasant to their present accumulated "knowledge" to date.

https://www.youtube.com/watch?v=2v5nAYU2GD0

WITHOUT ACCURATE OBSERVATIONS, "science" isn't even "science fiction". (Talking about "billions of years ago" IMMEDIATELY falls under the category of fantasy, imagination, fiction, NOT SCIENCE; unless you are claiming that you personally existed that long ago and are declaring your personal, honest, accurate observations) If you can't look at and acknowledge the facts, you are choosing to live in a delusionary state of your own making; SO PLEASE LOOK AT THE FACTS!

https://www.blastthetrumpet.org/PublicLetters/AAAUpdatedPublicAlertsMattersofLifeandDeath/Updates053016/Evidence%20of%20GOD.pdf - evidence of God.

If you spend the next week or so looking at ALL the evidence referenced in that note, I may be somewhat redundant here; but I am going to list SIGNIFICANT FACTS ABOUT WHAT WE DO OBSERVE; (understand I just plainly showed the whole world; from the largest database in the history of mankind, that the popular evolutionary paradigm taught in public education IS NOT OBSERVED. Clearly evolutionary theory has NO scientific support, IT IS NOT OBSERVED! and from its foundational premises throughout is nothing but imaginary fiction!), so here I am going to list WHAT WE DO OBSERVE and HOW IT PLAINLY SUPPORTS THE BIBLICAL ACCOUNT, EXCLUSIVELY! (By "exclusively" I mean that in all the world; among all literature of antiquity, there exists no record that can remotely compare to the vast knowledge contained in the Holy Bible, information that pertains to all scientific disciplines, information that pertains to all aspects of life and diverse professions, the largest compilation of historical information in ancient literature and confirmed by archeologists and anthropologists as accurate, information for governing the affairs of mankind, information about the scientific laws governing the universe, information that not only has been verified about our past but has accurately foretold detailed events in history on through current events happening before the eyes of the whole world; information that leads to the very SOURCE of fundamental TRUTH and ALL COMBINED Wisdom, Knowledge and Understanding in the Universe! - so some books and volumes of books may contain bits and pieces of information about what we observe and the process behind what we observe; but no other book comes close to the combined knowledge and unique attributes of the Record of our Creator; given to us all, as contained in the Holy Bible! -

https://www.blastthetrumpet.org/PublicLetters/AAAUpdatedPublicAlertsMattersofLifeandDeath/Updates053016/The%20Ineffably%20Awesome%20Holy%20Bible.pdf) So let's begin: THE OBSERVABLE SCIENTIFIC EVIDENCE THAT SUPPORTS THE BIBLICAL ACCOUNT OVERWHELMINGLY: A) FIRST WORDS WE READ IN THE FIRST VERSE! In the Beginning...

http://biblehub.com/genesis/1- 1.htm and http://biblehub.com/john/1-1.htm and

http://biblehub.com/john/1-3.htm and

http://biblehub.com/john/1-4.htm and http://biblehub.com/colossians/1-16.htm

https://www.youtube.com/watch?v=ZV0hUldrYp4 I could cite many more verses here but the Biblical Account plainly states that creation (the universe) had a Beginning. And science confirms this in the following ways:

1)

https://www.google.com/webhp?sourceid=chromeinstant&rlz=1C1GIGM_enUS535US535&ion=1&espv=2&ie=UTF8#q=mathematical%20proof%20that%20the%20universe%20has%20a%20beginning - In laymen's terms reverse extrapolation of presently observed expansion (regardless of steady state theory or big bang) means that IF present expansion is indicative of the past, then at one time energy and mass was combined in one location. Mathematics and Laws of Motion proves what I saw: https://www.youtube.com/watch?v=ri3R6vFVfEg&feature

2) "This argument about whether or not the universe had a beginning, persisted into the 19th and 20th centuries. It was conducted mainly on the basis of theology and philosophy, with little consideration of observational evidence. This may have been reasonable, given the notoriously unreliable character of cosmological observations, until fairly recently. The cosmologist, Sir Arthur Eddington, once said, 'Don't worry if your theory doesn't agree with the observations, because they are probably wrong.' But if your theory disagrees with the Second Law of Thermodynamics, it is in bad trouble. In

fact, the theory that the universe has existed forever is in serious difficulty with the Second Law of Thermodynamics. The Second Law, states that disorder always increases with time. Like the argument about human progress, it indicates that there must have been a beginning. Otherwise, the universe would be in a state of complete disorder by now, and everything would be at the same temperature. In an infinite and everlasting universe, every line of sight would end on the surface of a star. This would mean that the night sky would have been as bright as the surface of the Sun. The only way of avoiding this problem would be if, for some reason, the stars did not shine before a certain time. " - http://www.hawking.org.uk/the-beginning-of-time.html - Stephen Hawking in this paragraph actually expresses some rational thought, competent and coherent arguments, regardless of the rest of his statements in this citation. I want to point out a few things here for emphasis. The Second Law of Thermodynamics is in DIRECT OPPOSITION to the entire evolutionary constructs! They absolutely, positively CANNOT both be true. Evolution is claiming that order came out of chaos and the 2nd Law of Thermodynamics is saying that everything we observe breaks down, cools off, disintegrates, becomes more and more disordered over time. (The Word of GOD puts it this way - http://biblehub.com/2_corinthians/4-18.htm - EVERYTHING WE SEE (visible matter) IS TEMPORARY, had a beginning and will most definitely have an end. Our Creator tells us matter as it presently exists is temporal! finite! and the 2nd Law of Thermodynamics supports that).

So, the 2nd Law tells us; CONTRARY to evolution, that DNA over time will tend to lose, NOT gain, information, planets and stars will tend to DIE off, species will become extinct until there are none, etc. And not surprisingly this is exactly what is observed.

The scriptures also PLAINLY STATE that this creation is SUBJECT TO DECAY!

Future Glory

...**19**The creation waits in eager expectation for the revelation of the sons of God. **20**For the creation was subjected to futility, not by its own will, but because of the One who subjected it, in hope **21**that the creation itself will be set free from its bondage to decay and brought into the glorious freedom of the children of God.... https://biblehub.com/romans/8-21.htm

So, creation had a Beginning in which it was new and has been deteriorating ever since.

https://www.google.com/webhp?sourceid=chromeinstant&rlz=1C1GIGM_enUS535US535&ion=1&espv=2&ie=UTF8#q=more%20than%2098%25%20of%20all%20species%20have%20gone%20extinct Currently, we are witnessing mass extinction events worldwide; while these delusionary persons remain in psychological denial; still desperately clinging to their beloved theory in the vain hope that there is no GOD. Evolutionists INTENTIONALLY dismiss this scientific law BECAUSE it CLEARLY FALSIFIES what they believe! Consider what they THINK is scientific fact, (deep time of billions of years COMBINED with less ordered and complex life forms becoming more ordered and complex life forms (macro-evolution); they BELIEVE this DESPITE the fact that EVERYTHING we OBSERVE proves the OPPOSITE! But THEN THEY ACTUALLY ARE SO DECEIVED, THEY START POINTING AT "THE FOSSIL RECORD", as if the fossil record somehow supports evolutionary lunacy! evolutionists are SO deceived or self-deluded that they can't comprehend that the entire collection of fossils to date SOUNDLY FALSIFIES their religious evolutionary beliefs!

https://www.facebook.com/1mill.creationist/posts/872158116148696:0 LARGER AND MORE COMPLEX CREATURES ARE OBSERVED IN THE FOSSIL RECORD! http://jcourt.net/an-ecuadorian-giant25-feet-tall/ FAR MORE SPECIES THAN WE HAVE ON EARTH TODAY! (Scientists claim as much as a whopping 99% more! -

https://www.google.com/webhp?sourceid=chromeinstant&rlz=1C1GIGM_enUS535US535&ion=1&espv=2&ie=UTF8#q=99+percent+of+species+on+earth+extinct) hmmmm, their proposal is evolution causes BENEFICIAL changes (despite all observed natural mutations to date) over time, and they CLAIM billions of years have transpired (more than sufficient time to support their irrational claims) and YET instead of MORE SPECIES we have as little as 1% of all those that have ever existed on earth with MORE going extinct almost daily! -

https://www.google.com/search?q=extinction+events+currently+underway&rlz=1C1GIGM_enUS535US535&oq=extinction+events+currently+underway&aqs=chrome..69i57j69i60&sourceid=chrome&es_sm= 0&ie=UTF-8 and http://www.youtube.com/results?search_query=mass+die+offs+worldwide) but not only with the passage of time are the NUMBER of species dying off rapidly but the SIZE and OVERALL HEALTH OF ALL LIFE ON EARTH!

https://www.google.com/search?q=animals+humans+and+insects+were+larger+in+the+past&rlz=1C1GIGM_enUS535US535&oq=animals+humans+and+insects+were+larger+in+the+past&aqs=chrome..69i57j69i60&sourceid=chrome&es_sm=0&ie=UTF-8#q=Species+were+larger+in+the+past (the only repeatable experiment that I am aware of showing WHY species were larger and healthier in the past has been done by Christian Creationist, Dr. Carl Baugh - http://www.youtube.com/watch?v=N63lhtx2q8o&list=PL617ADBAD9F6F8790)

The cognitive dissonance in these persons is astounding! IF they get saved, it ABSOLUTELY will be PROOF of GOD; because they are presently impervious to reason, knowledge, science and facts; so, it most definitely will take a SUPERNATURAL ACT to reach them! (The evidence in support of this is how many cannot read or look at the actual evidence supporting the Biblical Account, and how many I have encountered who ACTIVELY DENY reality like the Real Mt. Sinai, The Flood, ashen remains of sodom and gomorrah and the hundreds of sites uncovered by archeologists, the recorded history documented by anthropologists, the fossil evidence of giants and much more ubiquitous REAL evidence that annihilate their ridiculous notions and worldviews.) In my honest opinion, the people I have encountered are showing that absolutely nothing short of a Divine Miracle can deliver these persons from such severe brainwashing. (The Word of GOD put it this way - http://biblehub.com/matthew/24-24.htm and http://biblehub.com/romans/1-28.htm and http://biblehub.com/romans/1-22.htm and http://biblehub.com/2_timothy/3-7.htm When people make idols out of "scientists" they are putting those fallible people on a pedestal that only God our Creator deserves, and is why they actually believe "life" was created by one of them in a laboratory, when none of them ever came close! Not only can they not face scientific facts, actual history, but they can't even look at the events transpiring in front of the eyes of the whole world proving the Biblical Account is Most Definitely Divinely Inspired.

https://www.blastthetrumpet.org/PublicLetters/AAAUpdatedPublicAlertsMattersofLifeandDeath/Updates053016/End%20Times%20Verse%20By%20Verse.pdf and

https://www.google.com/webhp?sourceid=chromeinstant&rlz=1C1GIGM_enUS535US535&ion=1&espv=2&ie=UTF8#q=mass+extinction+events+currently+underway+worldwide

https://www.youtube.com/results?search_query=mass+die+offs So the universe had a beginning, earth had a beginning, everything we can see had a beginning AND it ALL has an end. 2nd Law of Thermodynamics is clearly, plainly, observably TRUE and CONFIRMS THE BIBLICAL ACCOUNT; while evolutionary theories from cosmology through the theory of biological evolution (darwinism) are ALL clearly, plainly, observably FALSE; so FALSE, it is only religious desperation of those who desperately WANT there to be no God that keeps it from being discarded as the complete and total nonsense that it is! For millenniums the Holy Bible has declared a BEGINNING, even when steady state and an infinite universe was a popular concept and as Hawking put it; IF that were true, there would be no darkness we would have light everywhere. (By the way this is another argument FOR Intelligent Design and the Biblical Account of Creation) I have already covered some of the ADDITIONAL FACTUAL SUPPORT for the VERY FIRST VERSE; WHICH SHOWS THAT WHAT WE OBSERVE UBIQUITOUSLY (SCIENCE) CONFIRMS IT IN SO MANY WAYS IT IS ABSOLUTELY A SIGN OF MASSIVE IGNORANCE AND PERHAPS INSANITY TO DENY IT! In other words, not only does the physical reality of the universe confirm it (accurate observation and understanding of what is seen (can be perceived by your "natural" senses); but the spiritual reality behind it all. (accurate perception and understanding of what is unseen), MUST be spiritually discerned (is observed and understood accurately only by those who have been "born of God's Spirit"; not just born of the flesh - http://biblehub.com/1_corinthians/2-14.htm to REALLY understand the fullness of Creation EVERYONE MUST be "born again" -

https://www.blastthetrumpet.org/PublicLetters/AAAUpdatedPublicAlertsMattersofLifeandDeath/Updates053016/You%20MUST%20Be%20Born%20Again.pdf) GOD's WORD is TRUTH!

https://www.facebook.com/photo/?fbid=977229375689355&set=a.115635768515391 - Evidence of God - the physical universe (plainly observed reality overwhelmingly supports the Biblical Account throughout). Read the description for these memes, there plenty of information in them.

https://www.facebook.com/photo/?fbid=1380774058668216&set=a.115635768515391 - God is REAL! read the descriptions and notes following.

https://www.blastthetrumpet.org/PublicLetters/AAAUpdatedPublicAlertsMattersofLifeandDeath/Updates053016/NO%20MORE%20BRAINWASHING%20p1.pdf - observable reality supporting the Biblical Account and falsifying the nonsense of the popular evolutionary paradigm so many have been factually brainwashed to believe in.

https://www.blastthetrumpet.org/PublicLetters/AAAUpdatedPublicAlertsMattersofLifeandDeath/Updates053016/NO%20MORE%20BRAINWASHING%20p2.pdf - part 2

https://www.blastthetrumpet.org/PublicLetters/AAAUpdatedPublicAlertsMattersofLifeandDeath/Updates053016/The%20Ineffably%20Awesome%20Holy%20Bible.pdf - THE HOLY BIBLE IS FACTUALLY THE MOST AUTHORITATIVE

SCIENTIFIC REFERENCE IN THE ENTIRE WORLD AND HAS BEEN FOR MILLENNIUMS! no fallible "scientist" in all their many books have EVER published anything as astute, as accurate, as fundamentally true, as what is written in the Word of GOD! (especially virtually anything and everything any of them have postulated in modern theoretical physics) IN OTHER WORDS, the scientific findings in the Holy Bible contain NOT JUST ACCURATE OBSERVATIONS of past, present and even "FUTURE" events (thousands upon thousands of "predictions" that not only came to pass historically but are happening in front of the eyes of the whole world presently!) BUT unlike all the total and complete imaginary, facile twaddle published in the name of science by so many of these deceived and deluded "scientists" today, actually has NUMEROUS TESTABLE and REPEATABLE constructs THROUGHOUT! (I cover this in more detail; not only in my citations of other theses I have written like https://www.blastthetrumpet.org/PublicLetters/AAAUpdatedPublicAlertsMattersofLifeandDeath/Updates053016/The%20Sufferings%20of%20Christ.pdf , but as you read on) I make this statement because of how many of them REFUSE to even consider citing the most accurate reference work in the entire history of mankind as something that has been submitted for "peer review"; WHEN NOTHING ELSE THAT HAS EVER BEEN PUBLISHED IN THE ENTIRE HISTORY OF MANKIND HAS UNDERGONE MORE SCIENTIFIC SCRUTINY! (In other words, the Holy Bible is BY FAR the most reliable reference material in written format on our globe; BY FAR!!!; and yet these people revere the imaginations of obscure and provably deceived and self-deluded persons as their chosen support for "science". GET THIS THROUGH YOUR DENSE SKULLS YOU PROVABLY BRAINWASHED, RIDICULOUSLY STUBBORN PERSONS! (addressing the hard headed evolutionists and atheists) NOT ONE! NOT ONE! I repeat NOT ONE! "Peer review" you can cite holds even the tiniest spark of pervasive scientific scrutiny the contents of the Holy Bible have meticulously undergone FOR MILLENIUMS! As such, you can consider, some obscure publication reviewed by a handful of persons all funded by the same evil and wicked people to be your "evidence" if you want to; but the rest of SANE and REASONABLE humanity PREFER to cite the scientific findings that HAVE STOOD THE TEST OF TIME AND PRESENTED TO BILLIONS OF SOULS OPENLY ALL OVER THE WORLD; THROUGHOUT HISTORY TO THIS PRESENT DAY! As such, any "new" ideas that appear to defy this reference work, is FAR MORE LIKELY that those "new" ideas, ideologies are IN ERROR, than that they in anyway falsify the contents of the Holy Bible, that provably accurately records what we observe AND has had the greatest minds and most powerful persons in the History of the World unanimously acknowledge and refer to it themselves. (The fact that so many arrogant and ignorant evolutionists and atheists don't or choose to ignore its contents, would be yet more evidence of your factual brainwashing and self-delusions; not to mention having a very adverse affect on your arrogant presumption that evolutionists are "intelligent and reasonable" and your likewise brainwashed ideology that conversely Biblical

Creationists are dumb, stupid and base their worldview on "blind faith" without the use of either intelligence or reason.)

Are you present day evolutionists and atheists actually claiming superior intelligence and knowledge to these Biblical Creationists who founded the very fields of modern sciences and won Nobel prizes for their discoveries? -

https://www.blastthetrumpet.org/PublicLetters/AAAUpdatedPublicAlertsMattersofLifeandDeath/Updates053016/Fathers%20of%20the%20Sciences%20and%20Great%20Scientific%20Minds.pdf and

https://www.facebook.com/answersforatheists/photos/a.183264395187217.1073741827.181573948689595/333385456841776/?type=1&theater

http://www.oocities.org/promo777/quotesof.htm

To those of us who know better, evolutionists believing stories from people diggin' up bones as more reliable than the proven history in the Holy Bible; they're false claims that their STORY of the evolutionary paradigm is supported by science and the Biblical Account is just myth; WHEN THE CONVERSE IS TRUE, is actually showing the world these people are in overt https://www.blastthetrumpet.org/PublicLetters/AAAUpdatedPublicAlertsMattersofLifeandDeath/Updates053016/Denial%20of%20REALITY.pdf. None of the fiction of the evolutionary paradigm is supported by science or facts; while the contents of the Holy Bible have been thoroughly verified.

https://www.blastthetrumpet.org/PublicLetters/AAAUpdatedPublicAlertsMattersofLifeandDeath/Updates053016/Creation%20Itself%20is%20Overwhelming%20Evidence%20of%20our%20Eternal%20Creator.pdf I stated that the Word of GOD covers ALL of Creation, seen and unseen, all matter and energy, the physical realm (covered in my notes as cited throughout my written defenses of Truth (the bulk of my notes - http://www.blastthetrumpet.org/PublicLetters/): Primarily they are written to lead every soul that reads them into understanding that they can and should have a direct and personal relationship with our Creator, the One True GOD; now and forever and to help people not to present themselves as https://www.blastthetrumpet.org/PublicLetters/AAAUpdatedPublicAlertsMattersofLifeandDeath/Updates053016/Mentally%20Challenged.pdf by teaching the Truth!

https://www.blastthetrumpet.org/PublicLetters/AAAUpdatedPublicAlertsMattersofLifeandDeath/Updates053016/Creation%20V%20Evolution%20Use%20Your%20Intelligence%20to%20Recognize%20INTELLIGENCE.pdf

(http://www.blastthetrumpet.org/PublicLetters/AAAUpdatedPublicAlertsMa
ttersofLifeandDeath/MoreUpdates031315/Evidence%20of%20GOD%20our%20Creator.pdf) and

https://www.blastthetrumpet.org/PublicLetters/AAAUpdatedPublicAlertsMattersofLifeandDeath/Updates053016/Creation%20Itself%20is%20Overwhelming%20Evidence%20of%20our%20Eternal%20Creator.pdf

https://www.blastthetrumpet.org/PublicLetters/AAAUpdatedPublicAlertsMattersofLifeandDeath/Updates053016/The%20Sufferings%20of%20Christ.pdf) .

I also stated that the Words of our Creator, as recorded in the Holy Bible, covers the spiritual realm supporting and presenting the manifestation of the physical properties of the universe. (If you are only interested in the PHYSICAL EVIDENCE and not all the other EVIDENCE supporting not only the existence of our Eternal

Creator but showing who He is; then you are showing a pre-disposition to ignoring some of the most recent "scientific discoveries" and conclusions stating that there is in fact an unseen cause for what is seen and that unseen cause is VERY INTELLIGENT!) : Here is a quote from the father of quantum physics, Max Plank, a theoretical physicist who originated quantum theory, which won him the Nobel Prize in Physics in 1918. Quote: “As a man who has devoted his whole life to the most clear headed science, to the study of matter, I can tell you as a result of my research about atoms this much: There is no matter as such. All matter originates and exists only by virtue of a force which brings the particle of an atom to vibration and holds this most minute solar system of the atom together. We must assume behind this force the existence of a conscious and intelligent mind. This mind is the matrix of all matter.” end quote -

http://howtogetamiracle.com/chapter-5-quantumphysics-the-window-into-the-unseen-realm/ - that quote is scientific phraseology for what has been written and known by all reasonable persons for millenniums.

http://biblehub.com/hebrews/11- 3.htm and http://biblehub.com/colossians/1-16.htm and http://biblehub.com/colossians/1-17.htm

Understand "faith" is NOT blind belief; faith is actually creative energy that brings and presents imaginations into tangible reality! FAITH is the UNDERLYING ENERGY THAT HOLDS AND PRESENTS THE REALITY OF THE PHYSICAL UNIVERSE; SPECIFICALLY, it is the FAITH (CREATIVE ENERGY/CREATIVE POWER) of our INTELLIGENT CREATOR! FAITH(INTELLIGENT DIRECTED ENERGY) presents (brings into existence and holds together) VISIBLE CREATION! FAITH is NOT synonymous with "belief"! FAITH IS THAT CREATIVE FORCE; ALL THAT THOUGHT, PLANNING HOW YOU WILL CONSTRUCT AND MANIFEST YOUR IDEAS! FAITH is ACTIVE ENERGY; not mere "belief". The final manifestation of FAITH (INTELLIGENTLY DIRECTED CREATIVE FORCE) IS PHYSICAL REALITY -

http://biblehub.com/interlinear/hebrews/11-1.htm (energy is necessary to bring mass and matter into existence, energy is absolutely necessary to present mass/matter. Conversely, energy is in NO WAY dependent upon mass/matter. - Einstein's Theory of Special Relativity can be falsified scientifically in SO MANY WAYS that had I incarnated at the same time, I would have encouraged him in the strongest terms NOT to humiliate himself by presenting such incredibly foolish, insupportable notions! (BEFORE YOU think "but everybody knows..." LOOK AT THE EVIDENCE! -

https://www.blastthetrumpet.org/PublicLetters/AAAUpdatedPublicAlertsMattersofLifeandDeath/Updates053016/HOW%20TO%20KNOW%20TRUTH%20IN%20A%20WORLD%20OF%20LIES.pdf and

https://www.blastthetrumpet.org/PublicLetters/AAAUpdatedPublicAlertsMattersofLifeandDeath/Updates053016/Too%20Many%20are%20Turning%20Science%20Into%20a%20Religion.pdf)

The REALITY of the spiritual realm has been documented scientifically (NDEs, paranormal activities, ghosts, poltergeists, they have all kinds of names, but people have devoted decades of study, observations of these subjects now - MAKING AND RECORDING ACCURATE OBSERVATIONS is the first criterion of science and for establishing the scientific method! - keep in mind what is being taught in the name of science, regarding the entire evolutionary paradigm, HAS NOT EVEN BEEN OBSERVED! (doesn't even meet the first criterion of actual science!) So, for those who dismiss these findings, the rest of us know you are in favor of not supporting science or reason; but only on continuing to cling to chosen deceptions and delusions religiously. (evolutionists and atheists are the ones whose worldviews are FACTUALLY built upon "blind faith"; NOT the creationists).

https://www.google.com/webhp?sourceid=chromeinstant&rlz=1C1GIGM_enUS535US535&ion=1&espv=2&ie=UTF8#q=documented%20paranormal%20activity

https://www.google.com/webhp?sourceid=chromeinstant&rlz=1C1GIGM_enUS535US535&ion=1&espv=2&ie=UTF-8#q=the%20science%20behind%20NDEs

http://www.victorzammit.com/skeptics/challenge.html

https://www.google.com/webhp?sourceid=chromeinstant&rlz=1C1GIGM_enUS535US535&ion=1&espv=2&ie=UTF8#q=unexplained%20hauntings%20ghosts When ancient civilizations all over the world clearly mention the existence of the spiritual realm, gods, spirits, angels, demons, devils, etc.

https://www.google.com/search?q=ANCIENT+CIVILIZATIONS+CLAIM+KNOWLEDGE+FROM+GOD&rlz=1C1GIGM_enUS535US535&oq=ANCIENT+CIVILIZATIONS+CLAIM+KNOWLEDGE+FROM+GOD&aqs=chrome..69i57&sourceid=chrome&es_sm=93&ie=UTF-8

AND the most accurate Reference Book on the entire globe throughout all recorded history (the Holy Bible); it makes ANYONE who denies the existence of the spiritual realm look EXTREMELY ignorant; lacking massive bodies of knowledge http://biblehub.com/hosea/4-6.htm (knows next to nothing about the pervasive many scientific disciplines and how ALL of them combined OVERWHELMINGLY lead to our Creator, the One True GOD; from whom all Truth and Knowledge (Science) comes! http://biblehub.com/colossians/2-3.htm); they are not only seriously lacking vast amounts of knowledge about world history, but of observable reality. Just because you may not comprehend these things, does NOT entitle you to claim there is no evidence - http://biblehub.com/matthew/13-15.htm; when anyone who looks into these subjects, even briefly, can see so much evidence as to make ANYONE look totally foolish who would deny such. So, the following touches on just a few topics of How the Word of GOD is speaking SIMULTANEOUSLY about BOTH the physical reality in which we exist AND the spiritual reality in which we exist (the ACTUAL Intelligently Designed universe/CREATION in which we exist).

https://www.blastthetrumpet.org/PublicLetters/AAAUpdatedPublicAlertsMattersofLifeandDeath/Updates053016/Understanding%20the%20Holy%20Bible%20Part%201.pdf - Understanding the Holy Bible part 1 of 2 https://www.blastthetrumpet.org/PublicLetters/AAAUpdatedPublicAlertsMattersofLifeandDeath/Updates053016/Understanding%20the%20Holy%20Bible%20Part%202.pdf - part 2 of 2

https://www.blastthetrumpet.org/PublicLetters/AAAUpdatedPublicAlertsMattersofLifeandDeath/Updates053016/Commentary%20on%20Genesis%20Part%201.pdf - Commentary of Genesis part 1 of 5

https://www.blastthetrumpet.org/PublicLetters/AAAUpdatedPublicAlertsMattersofLifeandDeath/Updates053016/Commentary%20on%20Genesis%20Part%202.pdf - part 2 of 5

https://www.blastthetrumpet.org/PublicLetters/AAAUpdatedPublicAlertsMattersofLifeandDeath/Updates053016/Commentary%20on%20Genesis%20Part%203.pdf - part 3 of 5

https://www.blastthetrumpet.org/PublicLetters/AAAUpdatedPublicAlertsMattersofLifeandDeath/Updates053016/Co mmentary%20on%20Genesis%20Part%204.pdf - part 4 of 5

https://www.blastthetrumpet.org/PublicLetters/AAAUpdatedPublicAlertsMattersofLifeandDeath/Updates053016/Commentary%20on%20Genesis%20Part%205.pdf - part 5 of 5

https://www.blastthetrumpet.org/PublicLetters/AAAUpdatedPublicAlertsMattersofLifeandDeath/Updates053016/The%20Omniscience%20of%20Our%20Creator.pdf

B) http://biblehub.com/interlinear/genesis/1-2.htm - Here we see all modern evolutionary constructs of the many versions of the "big bang theory" are ALL false! While I have not heard and read about all the millions of imaginations regarding this supposedly scientific theory, I have encountered dozens of them; including as cited from university professors with post graduate degrees and ALL of them err with respect to this fact declared in the Holy Bible. They all claim that space and time "expanded" into existence from what amounts to nothing and then ridicule and denigrate anyone who asks for the cause of this most miraculous event. But here our Creator tells us things that EXPLAIN what we observe! He tells us that PRIOR to His Creation of Light that there existed empty space.

Understand that in His Creation of the "heavens and the earth" He plainly tells us that He brought Light into the existence of Space (He created formless earth, water and Space when He created "the heavens and the earth"; PRIOR to Light). He also tells us earth (which if you look at the root of that word is basically telling us that is matter/mass) in the darkness HAD NO FORM, HAD NOT BEEN SHAPED yet! These are KEYS of Knowledge about WHY the current "big bang" theories are flawed! We have Divine Instruction telling us that the void is actually empty space and in that empty space or void it was dark. So, by Define Definition darkness is related to a correct understanding of space! Space, darkness, void, and FORMLESS MATTER (earth and water) existed PRIOR to Created Light! This is a VERY IMPORTANT topic to grasp as to WHY we are able to have both observable Light and darkness presently! (And also, to understanding that space, darkness, void is NOT "nothing" but rather an emptiness of unspecified region; in which energy and matter was brought into perceivable existence.) I make this distinction because the big bang theories say that "nothing" existed and that space is a something created in the void of nothing when our Creator tells us void and space are essentially identical terms (an empty region of unspecified dimension of no mass or energy) and He further tells us that in the void of space earth and water was formless (mass, liquids, solids; the elements had not been shaped as yet to be perceived) Our Creator is telling us how He conceived to bring the forms of energy and mass into existence in the void of space. So here He tells us formless mass (what scientists are speculating on concerning "dark matter" and "dark energy" - I am NOT talking about their wild imaginations, I am attempting to bring them back to REALITY and Truth by telling everyone that "dark matter" is all mass in the universe that does not emit light, is NOT a source of light, regardless of size) existed prior to His Creation of Light. Space and Void is a region of emptiness in which everything else about our Created Universe was brought into existence! The presence of formless matter in space is one of the reasons, despite all the stars and galaxies, that we do not see constant visible light but can discern both day and night, light and darkness. THIS IS A KEY PRESENT-DAY OBSERVATION THAT CONFIRMS YET AGAIN THE BIBLICAL ACCOUNT; and does NOT support the current evolutionary constructs like the many constantly morphing theories and connotations of the "big bang". http://education.jlab.org/glossary/abund_uni.html - look at what WE OBSERVE the most abundant elements in space are formless earth and water! (in other words, what we see FORMED exists in the midst of what is unseen (not visible to the naked eye) elements that comprise primarily water!) " For this they willingly are ignorant of, that by the word of God the heavens were of old, and the earth standing out of the water and in the water: " -

http://biblehub.com/2_peter/3-5.htm Space is filled mostly of the ubiquitous presence of HYDROgen!

http://dictionary.reference.com/browse/hydro- hydro-1 ♣ Word Origin 1. a combining form meaning "water," used in the formation of compound words: hydroplane; hydrogen. Expand Also, especially before a vowel, hydr-1. OriginExpand hýdōr water Do you REALIZE THE SIGNIFICANCE?! FOR MILLENNIUMS, Our CREATOR HAS TOLD US EXACTLY HOW HE CREATED THE UNIVERSE! LONG, LONG, LONG before we scientifically observed what He has plainly told us; for all the world to read! Let's examine the Biblical Account further: "Let there be Light!" http://biblehub.com/genesis/1-3.htm - what scientific confirmation do we OBSERVE as EVIDENCE for the Biblical Account?

https://www.google.com/webhp?sourceid=chromeinstant&rlz=1C1GIGM_enUS535US535&ion=1&espv=2&ie=UTF-8#q=cosmic%20background%20radiation%20is%20the%20oldest%20light

NOTICE scientists call it the "oldest light"! Notice this is EMPIRICAL EVIDENCE for what has been written plainly for millenniums; FURTHER NOTE it was a recent discovery EVEN THOUGH the Word of GOD as recorded in the Holy Bible has told us all this plainly THOUSANDS OF YEARS BEFORE we even had the means of ascertaining it! Light separate from the darkness.

http://biblehub.com/genesis/1-4.htm - The universe most definitely had a beginning and according to rational scientists, they give reasons why that is true. While they claim that it is old, it cannot be so old that it overcomes their perplexity as to why we are able to perceive both light and darkness. From what appears to be an astrophysics professor from Berkley (Remember Stephen Hawking made a similar statement in the previous citation above): "* In spite of many, many stars and galaxies, the night sky is dark We do not know whether the Universe is infinite in size, but observations (the CMB, for instance) clearly indicate that it has a finite age, which the Big Bang theory and associated calculations set at 13.7 billion years. As a result, we can only see objects within 13.7 billion light years of us: the distance light could have traveled since the Big Bang. If the Universe were infinite in size and age, we would expect every line of sight to encounter the surface of a star, and the night sky to therefore be bright." -

https://www.google.com/url?sa=t&rct=j&q=&esrc=s&source=web&cd=1&cad=rja&uact=8&ved=0CCAQFjAA&url=http%3A%2F%2Fastro.berkeley.edu%2F~nmcc%2Fteaching%2FC10_Fall07%2FBig%2520Bang%2520soln.doc&ei=JutLVJW4OIS0yATbn4L4Dw&usg=AFQjCNF4wAMdK5HtRR0mJy9SUoX8lICdEw&sig2=Ne8dVJPuWQ_fgAJ93PKufA&bvm=bv.77880786,d.aWw

But here again our Creator tells us that the reason for WHAT WE OBSERVE is that He has deliberately separated Light and darkness!

The First Day: Light ...3Then God said, "Let there be light"; and there was light. 4God saw that the light was good; and God separated the light from the darkness. 5God called the light day, and the darkness He called night. And there was evening and there was morning, one day.

If He had not done this and designed our eyes to see only visible light as opposed to the entire Electromagnetic Spectrum, then we would not be able to distinguish colors or objects, but would be blinded by the inundation of all the pervasive cosmic background radiation and all the properties of Light that came into existence when He said, "Let there be Light!" This separation of Light and darkness; our ability to distinguish this separation, is rationally explained in the Biblical Account; while it baffles current evolutionary cosmologists the world over. Our Creator tells us that the reasons are: 1) formless mass (elements of earth and water) existed prior to the creation of Light in the void of space and 2) He intentionally separated Light from darkness and 3) He gave us perceptive abilities to discern as much (as we discover not only by our present observations but also by continuing to read our Creator's explanation about His Creation as found in the Holy Bible.) 5God called the light day, and the darkness He called night. And there was evening and there was morning, one day.

DIVINE DEFINTIONS are very important to note; because they help us understand not only what is written in the Holy Bible; but in acquiring truthful knowledge about all of His Creation. God not only separates(ed) Light and darkness, but calls all forms of Light, Day, (physical and spiritual aspects of Light is called Day) and likewise, physical and spiritual aspects of darkness He calls night. He further defines the distinction between Day and night as morning and evening. This is WHY ALL generations; ALL He has made is all but ONE GREAT DAY before our Creator. (One Epoch in which all His Creative Thought about His Creation is made visible and He tells us about this in His Word, the Holy Bible, that covers all matters of great importance past, present and future). So PRIOR to Christ, ignorance and darkness was upon the waters (people - http://biblehub.com/revelation/17-15.htm) and then the Day Star Arose - http://biblehub.com/2_peter/1-19.htm,

Christ, the Sun of Righteousness - http://biblehub.com/malachi/4-2.htm,

the Bright Morning Star - http://biblehub.com/revelation/22- 16.htm

DAWNED (Morning, Day, Light came to humanity, evening, night, darkness is fading away!) So, from the BEGINNING of visible Creation; until it is transformed from being temporal and made eternal is one EVENING of ignorance in which the vast amount of humanity did not KNOW our Creator, then His Incarnation, began to shine Light on earth in which the darkness of ignorance is fading away and leads ultimately to the enlightenment of every man, woman and child on earth! (no more darkness or night on earth whatsoever) -

http://biblehub.com/jeremiah/31-34.htm

and http://biblehub.com/revelation/22-5.htm

The Destroyers of Jerusalem Destroyed ...5You will flee by the valley of My mountains, for the valley of the mountains will reach to Azel; yes, you will flee just as you fled before the earthquake in the days of Uzziah king of Judah. Then the LORD, my God, will come, and all the holy ones with Him! 6In that day there will be no light; the luminaries will dwindle. 7For it will be a unique day which is known to the LORD, neither day nor night, but it will come about that at evening time there will be light.... Ultimately, it is His Presence and that of all the starry host with Him (angels, the Redeemed) that does away with all darkness, once and for all upon His Glorious Return!

http://biblehub.com/2_thessalonians/1-8.htm Christ's Coming ...7and to give relief to you who are afflicted and to us as well when the Lord Jesus will be revealed from heaven with His mighty angels in flaming fire, 8dealing out retribution to those who do not know God and to those who do not obey the gospel of our Lord Jesus.9These will pay the penalty of eternal destruction, away from the presence of the Lord and from the glory of His power,... http://biblehub.com/2_peter/3-10.htm

The Day of the Lord 10But the day of the Lord will come like a thief, in which the heavens will pass away with a roar and the elements will be destroyed with intense heat, and the earth and its works will be burned up. 11Since all these things are to be destroyed in this way, what sort of people ought you to be in holy conduct and godliness,... This is why all the generations of incarnate life on earth is referred to as ONE DAY. (a time of ignorance, followed by a time of enlightenment) - http://biblehub.com/acts/17-30.htm

Paul Before the Areopagus ...29"Being then the children of God, we ought not to think that the Divine Nature is like gold or silver or stone, an image formed by the art and thought of man.30"Therefore having overlooked the times of ignorance, God is now declaring to men that all people everywhere should repent,31because He has fixed a day in which He will judge the world in righteousness through a Man whom He has appointed, having furnished proof to all men by raising Him from the dead."... When those who would search out the mysteries contained in the Scriptures realize that our Eternal Creator SIMULTANEOUSLY is TEACHING us all not just about VISIBLE CREATION (the physical universe with the "natural laws" and properties) but also about His INVISIBLE CREATION (that the physical universe is presented by the supernatural spiritual realm) THEN the reader begins to see the meanings of the Dreams, Visions, Revelations and Prophecies within the Scriptures more clearly and better grasps the Reality in which we all exist is NOT just physical matter and "natural laws" but spiritual ones; supernatural presence, existing apart from and yet within and all around what is visible and observable.

As such GOD is SIMULTANEOUSLY telling us how to define not only the 24-hour periods or evenings and mornings people think of as days on earth, but how to recognize the various epochs on earth. Thus, in one part of the Scriptures we understand that there is ONE GREAT DAY in which all of Genesis through Revelation is fulfilled, but we also understand that from the time Christ Incarnated to His Glorious Return is another Great Day. And we further understand dark empires on earth each signified times of night or evening on earth, etc. such that then the prophecies and visions correlate directly with observed history as I point out in this timeline below and in previously cited notes regarding the spiritual connotations and fulfillment of the Words of our Creator.

https://www.blastthetrumpet.org/PublicLetters/AAAUpdatedPublicAlertsMattersofLifeandDeath/Updates053016/CHRIST%20IS%20THE%20FULFILLMENT%20OF%20THE%20LAW.pdf - Christ is the fulfillment of the Law and Prophecies.

http://www.godempowersyou.com/documentation/HistoricalTestimonyandMinistry/MansChronologicalExistenceExplained.pdf - this explains the SEVEN GREAT DAYS and the Seven Days within the SEVENTH GREAT DAY.

http://biblehub.com/isaiah/30-26.htm

https://www.youtube.com/watch?v=ZV0hUldrYp4 - cycles of sevens and other cycles not just in the BIBLE but in CREATION is ONE OF MANY "SIGNATURES of GOD" our Eternal Creator: the presence of accurate calendars, that fit Biblical time frames, the construction of constellations and our solar system, fitting the Biblical Account precisely corresponding not only to the distant past but recorded history AND even encrypted messages WITHIN the accurate history and prophecies recorded in the HOLY

BIBLE, ALL COMBINED ARE OVERWHELMING, IRREFUTABLE EVIDENCE of the DIVINE INSPIRATION of the HOLY BIBLE.

https://www.youtube.com/watch?v=ZV0hUldrYp4

https://www.youtube.com/results?search_query=bible+codes+correlate+to+modern+events

https://www.youtube.com/watch?v=al9cN9s7MP0

https://www.youtube.com/watch?v=OeEPcDaHicl&list=PLbN27mU6VgwUcOJX8XK7QeuQvLe_CNL1 -

I cite this just to give a SMALL EXAMPLE of the Divine Encryption encoded in the HOLY BIBLE that took the advent of super computers to even begin to understand HOW NUMEROUS, COMPLEX and SPECIFIC are these revelations embedded in the Divinely Inspired Holy Writ!!!!!!!! I point this correlation out between the Biblical Account and the recorded history of mankind; that the Biblical account of Creation is supported by the sciences, including world history to this present day as these cited expositions mention. In supporting the Biblical Account, I mention them also because they falsify the nonsense of evolutionary constructs as forced on children in public education today. (not only does the physical universe falsify what evolutionists and atheists believe, but all of history does as well!) Conversely, the sciences and history directly correlate with the Biblical Account! (Creationists base their worldview on science and reason; NOT the evil-u-shunists and atheists!) I find it astonishing that anyone would question the Biblical Account of the Creation in which we all exist, particularly when even unbelieving scientists acknowledge mankind (physical attributes are common to soil, dirt, dust, earth) - https://www.youtube.com/watch?v=5xVHzM08JG0.

Notice the presenter is honest enough to admit he doesn't know our Eternal Creator. Understand that if you do NOT KNOW our CREATOR, you CANNOT be CERTAIN, absolutely CERTAIN, of ANYTHING! What are you to judge it against? popular opinion? someone else's confirmation? someone you esteem as more intelligent? WHO OR WHAT CONFIRMS WHAT YOU THINK YOU KNOW AS CERTAIN FACT?! If you say the sky is blue, how do you know? how do you know it isn't really red? I have in other notes cited above pointed out that we all perceive "reality" in very subjective and even radically different ways! (Examples are:

https://www.google.com/webhp?sourceid=chromeinstant&rlz=1C1GIGM_enUS535US535&ion=1&espv=2&ie=UTF-8#q=synthesia+and+savants, color blind, blind, deaf, or many other persons who do not or cannot perceive reality as do many of us) None of us can KNOW for certain what "blue" is to others (their connotations in their mind); likewise, EVERYTHING you have been taught is SUBJECTED to NOT ONLY YOUR METHOD OF PERCEPTION(s); BUT THOSE WHO PASSED ON THEIR OPINIONS TO YOU. (Everything you THINK you know is HIGHLY SUBJECTIVE) and if you do NOT KNOW our CREATOR, you have NO ABSOLUTE AUTHORITY (THERE IS NO PERFECT AND CERTAIN TRUTH WITHOUT GOD) to run those HIGHLY SUBJECTIVE perceptions and opinions by. So instead "scientists" run their subjective perceptions and opinions by other equally highly subjective and fallible persons. So, ANYTHING someone who does NOT KNOW GOD espouses is MORE THAN LIKELY FALSE. For example, looking at a variety of bones of different species and then thinking "dinosaurs 'evolved' into birds", as opposed to just realizing those are distinct species that died, no one has any proof whatsoever of whether they had offspring, let alone evolved into something else. For example, we have ostriches, emus, chickens, swans, eagles, all birds with different sizes and abilities, including some that fly and some that don't. Likewise, we have fish, amphibians, reptiles, and if all those were dead today RIGHT NOW, evolutionists would dig up their bones and try to claim that EVEN THOUGH THEY LIVED AND DIED AT THE EXACT SAME TIME, that those DIFFERENT species morphed or "evolved" from either legs/fins to no legs/fins or vice versa. THEY DO THIS ALL THE TIME WITH BURIED BONES THEY DIG UP! (Starting with an impossible assumed paradigm of fictional/imaginary constructs, they dig up bones and then purposefully place the bones of different species in a desired order in a vain attempt to CLAIM that one kind of species evolved into another.) IT ISN'T TRUE TODAY, and it isn't true in the past! Notice that those who DO KNOW GOD (not those who CLAIM to Know, but those who PROVABLY KNOW the ONE TRUE GOD), universally confirm the Biblical Account! So, someone who admits he does not know is attempting to debate someone who does. As such, the one who admits he does not know SHOULD learn from the one who does, but the reason he does not do so is that he THINKS other people do not KNOW our Creator, (chooses to think ALL those claiming so are LYING). But I cited the presentation because he is a biologist who can see Creation must be Engineered and Designed; only he then thinks that it was only done so INITIALLY, instead of pervasively and continuously. I find it astonishing he can see programming, engineering, and yet still somehow be in doubt as to the existence of the Engineer!

http://biblehub.com/genesis/1-6.htm - 6Then God said, "Let there be an expanse in the midst of the waters, and let it separate the waters from the waters."

Creationists and others present ideas about the expanse or firmament.

https://www.youtube.com/watch?v=sZeDai1mPDM and Dr. Carl Baugh has conducted an experiment in an attempt to offer supporting evidence of his theory.

http://www.keelynet.com/biology/baugh.htm

The LORD taught me personally more about the spiritual connotations of His Word, than the physical; so, my comments about how the physical evidence in observable creation, the visible universe and how it relates to the Biblical Narrative, is how I personally comprehend what I am reading. Also, I tend to recall the dream I had where I saw the Lord take all the elements, matter, mass in the universe and combine them as an enormous superheated sphere and then spin it. I noted that the globules that went flying off the giant mass would stretch like lava and yet while exploding would form spherical droplets and dissipate into smaller and smaller particulates; began immediately after it was all set in motion. All the superheated elements began cooling and breaking down as soon as the mass started moving through the darkness in the void of space. So, we have observed cosmic background radiation and rather formless water as elements like pervasive hydrogen in space; but if the Lord had not set boundaries or made the expanse then entropy would result in all mass reaching a state of equilibrium (dissipation down to quarks ubiquitously at an alarming rate.) In the void of space there are forms of energy that we have observed and even though scientists call them "natural" all forms of energy and matter are anything but "natural" in their fundamental essence. (not just the origin and initial cause but even the principles that define forms of energy are supernatural) As such, scientists are trying to define what they call "dark energy" that seems enigmatic to them. Based on my reading of the text and what I saw in dreams and visions, I am persuaded at present that the "expanse" is one of the factors that causes the liquids, solids, and gases to hold together and form lakes of fire in space; stones of fire in space; otherwise known as stars and planets. In other words, I am convinced that it is MORE than just "gravity" that is responsible for the formation of stars and planets, galaxies, but that GOD our Creator has made something in the void of space that He calls the expanse or "firmament". Based on my comprehension of what I read that it is more than just our atmosphere, and more than just layers in the mantel of the earth that separates subterranean oceans

https://www.google.com/webhp?sourceid=chromeinstant&rlz=1C1GIGM_enUS535US535&ion=1&espv=2&ie=UTF8#q=subterranean%20ocean%20discovered from surface water and water in our air. But that there is an energy that actually fights against the tendency of mass and matter to break apart (something that resists entropy, such that even though creation is decaying, it is decaying at a rate that God Himself approves of, one that doesn't mean we labor in vain, like attempting to build something that only falls into dust by nightfall); an energy that is not the gravity within mass and matter but is all around visible mass and matter keeping it from disintegrating. And that "expanse" or "firmament" is part of what is holding stellar bodies (and all visible matter) as an observable form rather than completely and rapidly dissipating into the void of space altogether. (God tells us it is He that is holding creation before us all together. Col 1:15-19)

https://www.youtube.com/watch?v=TdVB3Pqwcdw - MUCH closer to observable reality, actual science! So, as I read the Biblical Account, I see that there were formless elements pervasive in the void of space and God added to those formless elements, which were mostly what comprises water molecules, energy in the form of light and the vibrations of His Spoken Word through it all, sound frequencies. So, space was filled with formless matter elements, energy, and vibrations that initiated all that we presently behold. And in the process, He began to form the mass and matter by forcefully separating the pervasive formless elements such that it was all no longer pervasive invisible formless mass and energy all vibrating in the void of space, but began to separate and concentrate (started to take form). Let us consider examples of how gravity cannot be the sole explanation for planetary and stellar formations, or the sole explanation for why they have not virtually instantaneously disintegrated into invisible particulates and indistinguishable forms of energy. Consider the plant fibers that make up paper. Look at the process.

https://www.youtube.com/watch?v=E4C3X26dxbM and the archaic method that anyone can do.

https://www.youtube.com/watch?v=UlX87tvei7c
Particularly note how that saw dust that would just blow away in the wind and further decompose and disintegrate

http://westernpulp.com/sustainability/westerns-containers and

http://www.gardening.cornell.edu/factsheets/orgmatter/ and

http://www.motherearthnews.com/organic-gardening/building-garden-soil-wood-mulchzmaz10onzraw.aspx#axzz3HwRGz7gg and http://ohioline.osu.edu/hyg-fact/1000/1083.html

There are ways to hasten the "natural" process, but as you can see wood and plant fibers break down into soil in very short time spans depending on the environmental conditions. Compare all that to what is necessary to forcefully hold those fibers together in order to prevent rapid decomposition and entropy from continuing. Applied chemical, heat and mechanical energy and massive amounts of it! Ultimately PRESSURE, APPLIED EXTERNAL PRESSURE, compacts the fibers together. It is NOT their "gravitational attraction" due to their mass holding them together at all! Likewise, dissimilar materials (matter/mass/elements) all over our planet are formed or held together by EXTERNAL PRESSURE primarily, NOT "gravity", exclusively, or even as the predominant cause! Look at sand under a microscope:

https://www.google.com/search?q=sand+under+microscope+snopes&rlz=1C1GIGM_enUS535US535&espv=2&biw=1066&bih=702&tbm=isch&tbo=u&source=univ&sa=X&ei=XIpWVLnaKNavyATCnoGAAg&ved=0CC8QsAQ&dpr=1.5

All those magnified components that make up sand makes you want to grab a handful and shout "I'm rich!" doesn't it. But I want you to notice how dissimilar all the particulates are and further notice how those grains of sand are NOT pulling themselves together by "gravity" within them; one grain to another is not fusing themselves together in your hand but rather friction and other causes of entropy further erodes them down into smaller and smaller particles. Yet WITH pressure, sedimentary and even metamorphic rock is formed from such dissimilar particles. Some would claim that pressure is primarily due to the gravitational attraction of a very dense core as opposed to the presence of layers of mass and sediment pushing against lower buried layers of particles and "atmospheric pressure".

https://www.google.com/webhp?sourceid=chromeinstant&rlz=1C1GIGM_enUS535US535&ion=1&espv=2&ie=UTF-8#q=atmospheric%20pressure Many would STILL attribute even "atmospheric pressure" to "gravity". But again, I ask those proposing and attributing to "gravity" as the cause, to not only consider the many examples that it takes EXTERNAL PRESSURE to fuse mass and matter together (even modern fusion reactions -

https://www.google.com/webhp?sourceid=chromeinstant&rlz=1C1GIGM_enUS535US535&ion=1&espv=2&ie=UTF8#

q=modern+theoretical+fusion+reactor+models+achieve+fusion+by+applying+external+energy) And then ask yourselves IF GRAVITY in MASS is the sole force holding matter together; then what is causing universal expansion? What is causing the galaxies and stars within galaxies to drift further and further apart? (why aren't ALL galaxies collapsing upon themselves as gravity pulls all the mass and matter together? That is, if gravity is the force alone, defying entropy.) Furthermore, what keeps the mass and matter within each star and planet from continuing to draw the matter all around it closer and closer together due to gravity until it becomes a tiny dense particle? God has provided that "gravity" is just negligible enough to allow common laws of motion to counteract it and other forms of energy. He tells us in His Word that by Him do all things consist. Col 1:15-19

https://www.google.com/webhp?sourceid=chromeinstant&rlz=1C1GIGM_enUS535US535&ion=1&espv=2&ie=UTF8#q=All%20mass%20and%20matter%20is%20more%20than%2098%20percent%20space each atom that comprises each element that comprises all the visible mass and matter in the universe has huge amounts of space between even the quarks. So, what APPEARS to be solid is mostly nebulous space with tiny amounts of particulates (just look at the pile of ash that is left after cremation). Apart from APPLIED EXTERNAL ENERGY mass, all matter, disintegrates until it is so small and so indistinguishable from the void of space that it becomes virtually imperceptible; apart from EXTERNAL APPLIED ENERGY all mass and matter goes through entropy and DOES SO RAPIDLY! (Especially, when viewed from concepts of infinity and eternity - SO REMEMBER THIS POINT WHEN ANYONE TRIES TO CLAIM FOSSILS support the evolutionary paradigm; instead of the Biblical Account of the Global Flood! For soft tissue, for plants, for all the single celled fossils to be preserved in the abundance we have IT HAD to be QUICK, it HAD to have PRESSURE and COVERING by sediments in order for them to exist at all!) So, when we look up at the night sky and can distinguish Light from darkness; concentrations of mass and matter that coincide with historic events on earth -

https://www.youtube.com/watch?v=BJ6szvfApNM and

https://www.youtube.com/watch?v=uGuQmfNNSQ&list=PLv7j4CHggW52D06EDjeAzP6qc2AHFzW0u and https://www.youtube.com/results?search_query=blood+moons+decoding+the+imminent+heavenl y+signs

and http://www.bethlehemstar.net/ ,

AND then consider that WHAT WE OBSERVE about forms of "natural" energy and mass; the universe, ALL OF

CREATION MUST BE CAREFULLY CONTROLED AND BALANCED or we get disasters like TOO MUCH PRESSURE would cause collapses of atomic structures and galaxies, stars, planets wouldn't even be able to exist or TOO LITTLE PRESSURE and it all would fly apart in virtually instantaneous entropy, disintegrating into the void of space into a state of equilibrium where all energy and mass would become indistinguishable and indifferent. So NOT ONLY MUST the "expanse", the "firmament" exist aiding in the formation of mass and matter, BUT it MUST be INTELLIGENTLY CONTROLLED! Such balances in our universe are what are commonly known as "fine tuning" arguments for Intelligent Design.

http://www.reasons.org/articles/fine-tuning-for-life-in-theuniverse

(In other words, what we observe IRREFUTABLY PROVES THAT GOD IS NOT ONLY IN ALL THINGS HOLDING THEM TOGETHER AND PRESENTING IT ALL TO US AS OBSERVABLE "REALITY"; BUT THAT HE IS ALL AROUND VISIBLE CREATION EXTERNALLY AS WELL! http://biblehub.com/ephesians/4- 6.htm and http://biblehub.com/colossians/1-17.htm

So, GOD, and His Intelligent Creative Power, PROVABLY has brought EVERYTHING into existence, BY HIS WORD, but not only MUST our Intelligent Eternal Creator, be the CAUSE and Initiator of EVERYTHING we observe in the entire universe; BUT HE MUST BE CONSTANTLY PRESENT WITHIN ALL THINGS AND SURROUNDING ALL THINGS in order for ANY of it to exist at all!!!!!!!!

https://www.youtube.com/watch?v=GRkzfs4_PSk

http://biblehub.com/acts/17-28.htm 27that they would seek God, if perhaps they might grope for Him and find Him, though He is not far from each one of us; 28for in Him we live and move and exist, as even some of your own poets have said, 'For we also are His children.'29"Being then the children of God, we ought not to think that the Divine Nature is like gold or silver or stone, an image formed by the art and thought of man.... and http://biblehub.com/ephesians/4-6.htm one God and Father of all who is over all and through all and in all. http://biblehub.com/genesis/1-7.htm God made the expanse, and separated the waters which were below the expanse from the waters which were above the expanse; and it was so. 8God called the expanse heaven. And there was evening and there was morning, a second day. Again, the Divine Revelations I received dealt more with how His Word applies to the separation of those who choose to do evil from those who choose to do good throughout the written record of the canonized Holy Bible and that Genesis is the Introduction of that Process and Revelation is the Emphatic Recap of that Separation Process.

https://www.blastthetrumpet.org/PublicLetters/AAAUpdatedPublicAlertsMattersofLifeandDeath/Updates053016/Commentary%20on%20Genesis%20Part%201.pdf But BASED ON MY READING COMPREHENSION of the text that what is commonly referred to as "outer space" or "celestial" as opposed to "terrestrial" our Creator refers to that space above in which the stars exist as heaven. On a side note, there are many connotations of the word "heaven"; in the physical and spiritual categories of that word. Some think of "heaven" as being "paradise", http://biblehub.com/luke/23-43.htm The Crucifixion ...42And he was saying, "Jesus, remember me when You come in Your kingdom!" 43And He said to him, "Truly I say to you, today you shall be with Me in Paradise." http://biblehub.com/2_corinthians/12-2.htm Paul's Revelation 1Boasting is necessary, though it is not profitable; but I will go on to visions and revelations of the Lord.2I know a man in Christ who fourteen years ago-- whether in the body I do not know, or out of the body I do not know, God knows-- such a man was caught up to the third heaven. 3And I know how such a man-- whether in the body or apart from the body I do not know, God knows-- 4was caught up into Paradise and heard inexpressible words, which a man is not permitted to speak. others think of heaven as a celestial supernatural city http://biblehub.com/revelation/21-1.htm A New Heaven and a New Earth 1Then I saw a new heaven and a new earth; for the first heaven and the first earth passed away, and there is no longer any sea. 2And I saw the holy city, new Jerusalem, coming down out of heaven from God, made ready as a bride adorned for her husband....

Others think of heaven in more mundane terms, such as everything outside of earth; beyond earth's atmosphere. However, if there is anything I have learned from our Creator is that His WORD affects and pertains to ALL His Creation, all things visible and invisible, all things physical and spiritual! And as such all of these connotations combined are like the tiny neurons in our brains that work together to give us a much larger picture of our perception of "reality". The scriptures are of no private interpretation http://biblehub.com/2_peter/1-20.htm from any of us but are IN DEPTH DIVINE REVELATIONS of layers upon layers of DIVINE Wisdom, Knowledge and Understanding!

http://biblehub.com/hebrews/4-12.htm and

http://biblehub.com/romans/11- 33.htm and

https://www.blastthetrumpet.org/PublicLetters/AAAUpdatedPublicAlertsMattersofLifeandDeath/Updates053016/The%20Omniscience%20of%20Our%20Creator.pdf And each one of us offering our perceptions on them, if so, we are in Christ and have received Him within our souls, are but one aspect or like one tiny neuron in the Mind of GOD!

https://www.blastthetrumpet.org/PublicLetters/AAAUpdatedPublicAlertsMattersofLifeandDeath/Updates053016/You%20MUST%20Be%20Born%20Again.pdf - physical creation indicates spiritual truth, so just as you had a birth into it, when you become spiritual born of Jesus Christ, your understanding of our universe increases like being born into a whole new reality that you weren't aware of previously. God gives you His Holy Spirit and your awareness of Creation and our Creator becomes real and magnified. Whereas you couldn't hear Him before, being born again causes you to hear God and know that He is OUR FATHER! Rom 8:7-15

And each of those views and perceptions COMBINED bring us closer to TRUTH and a right Understanding of both our Creator and His Creation. (Please stop fighting, debating and bickering among yourselves on who is correct or incorrect in their interpretations, IF YOU ARE ALL STANDING ON THE COMMON GROUND OF SALVATION IN CHRIST ALONE! -

http://biblehub.com/2_timothy/2-23.htm; most of the debates I have witnessed are arguments of ignorance to a greater or lesser extent. Some people choose to talk about and argue about things they really know nothing or very little about (even if they THINK otherwise); rather than just focusing on what is MOST IMPORTANT and that is that EVERYONE NEEDS TO KNOW THAT THEY CAN KNOW AND COMMUNICATE WITH OUR ETERNAL CREATOR PERSONALLY HERE AND NOW AND FOREVER!!!!!!!! Jesus Christ is the Teacher of us all. Jn 13:13) PLEASE look to all the souls still in darkness, who still do not personally KNOW our CREATOR! And spend your time and effort on introducing them all to the ONE TRUE GOD!!!!!!!!

https://www.blastthetrumpet.org/PublicLetters/AAAUpdatedPublicAlertsMattersofLifeandDeath/Updates053016/Soldiers%20of%20Christ%20First%20General%20Order.pdf (rather than bickering over whose private interpretation of Scriptures is the correct one; LET GOD BE THE ONE TEACHER OF US ALL! -

http://biblehub.com/matthew/23-10.htm and

http://biblehub.com/matthew/23-8.htm and

http://biblehub.com/john/14-26.htm and

https://www.blastthetrumpet.org/PublicLetters/AAAUpdatedPublicAlertsMattersofLifeandDeath/Updates053016/Are%20You%20A%20Christian.pdf) Only when we ALL personally KNOW and LEARN directly from our Creator, will He bring us all together in peace and harmony and see His Creation as He wills; without deceptions or delusions, and thereby see "eye to eye" concerning our Creator and His Creation.

http://biblehub.com/hebrews/8-11.htm and

http://biblehub.com/ephesians/4-13.htm

Those who have walked with Christ longer and experienced His Power, His Divine Revelations and Personal Mentoring, only help "younger" brothers and sisters in the Faith grow into such personal relationship and maturity in Christ; so that by His Grace these Messengers of GOD preserve unity among the Body of Christ and repair divisions among us that come in for so many reasons; as it is written: Unity in the Body 11And He gave some as apostles, and some as prophets, and some as evangelists, and some as pastors and teachers,12for the equipping of the saints for the work of service, to the building up of the body of Christ;13until we all attain to the unity of the faith, and of the knowledge of the Son of God, to a mature man, to the measure of the stature which belongs to the fullness of Christ.14As a result, we are no longer to be children, tossed here and there by waves and carried about by every wind of doctrine, by the trickery of men, by craftiness in deceitful scheming;...

What we presently observe clearly verifies the Biblical account of Creation, word for word; both regarding the physical reality of our universe and the spiritual. So let us continue.

http://biblehub.com/genesis/1-9.htm

The Third Day: Dry Ground 9Then God said, "Let the waters below the heavens be gathered into one place, and let the dry land appear"; and it was so. 10God called the dry land earth, and the gathering of the waters He called seas; and God saw that it was good. 11Then God said, "Let the earth sprout vegetation: plants yielding seed, and fruit trees on the earth bearing fruit after their kind with seed in them"; and it was so.12The earth brought forth vegetation, plants yielding seed after their kind, and trees bearing fruit with seed in them, after their kind; and God saw that it was good. 13There was evening and there was morning, a third day.

What I saw regarding the "days" of Creation was not like anything I have ever heard from anyone else or read anywhere else; but was an entirely different connotation. That all the "days" of creation are similar to the imaginations in your own mind when you set about any extensive project. Again, I will refer to building a house. In the process, one imagines what kind of house they will build, dimensions, size, colors, and then goes about refining the details of the overall vision as to what materials, what structural elements in each of the aspects or parts of the whole house. Detailed plans are drafted BEFORE even laying the foundation! Materials are gathered and then step by step ordered processes occur to MANIFEST or MAKE VISIBLE what was CREATED and IMAGINED in your mind! Creation is no different. Our CREATOR CREATED or IMAGINED EVERYTHING that is MADE MANIFEST in the universe over what we think of as the passage of time. So, there is no contradiction as some would claim in the CREATION account and the then following details of HOW what our CREATOR IMAGINED was then MADE MANIFEST.

http://biblehub.com/genesis/2-5.htm

http://biblehub.com/interlinear/genesis/2-5.htm and

http://biblehub.com/interlinear/genesis/2- 1.htm - notice that our Creator Imagined or Thought of ALL of CREATION; BEFORE making it VISIBLE or MANIFEST! The Creation of Man 1Thus the heavens and the earth were completed, and all their hosts.2By the seventh day God completed His work which He had done, and He rested on the seventh day from all His work which He had done. 3Then God blessed the seventh day and sanctified it, because in it He rested from all His work which God had created and made.4This is the account of the heavens and the earth when they were created, in the day that the LORD God made earth and heaven. 5Now no shrub of the field was yet in the earth, and no plant of the field had yet sprouted, for the LORD God had not sent rain upon the earth, and there was no man to cultivate the ground. 6But a mist used to rise from the earth and water the whole surface of the ground.... So, in the Revelations I had, The Seven Days were Divided into Categories that HOUSED THE ASPECTS of EACH OF "Days of Creation". And OVER OUR CONCEPT of TIME, bits of EACH OF THE SEVEN DAYS would be MADE MANIFEST. So, to grasp what "till ALL is FULFILLED" means

http://biblehub.com/luke/24- 44.htm

One has to LOOK AT ALL OF CREATION, ALL OF HISTORY, and SEE each plant coming into existence, each person, each animal as ALL part of the seven days of Creation. (Genesis Ch. 1 is like a Table of Contents for ALL that we observe about the entire universe, about our CREATOR and His Creation that has been MADE over "time".) Likewise, understanding the Book of Revelation. Since the Beginning, how many souls have strayed away from our Creator? How many have fallen? How many have become darkened? How many have thereby made themselves shipwrecked in the Sea of Humanity?

http://biblehub.com/1_timothy/1-19.htm and

http://biblehub.com/revelation/17-15.htm and thus trees that should have been for life http://biblehub.com/luke/6-44.htm; instead by receiving knowledge of how to do evil things from the wicked imaginations of the devil, the spirit of error, the spirit of antichrist, became dead and only fit to be burned http://biblehub.com/jude/1-12.htm and

http://biblehub.com/matthew/7-19.htm

When one considers all the forest fires that have ever occurred AND how many have by choosing to do evil deeds instead of repenting made their lives nothing but what should be destroyed (those who insist on raping, robbing, murdering others, rather than working peacefully together with all for what makes for life, and happy existence), are what the Book of Revelation and Jude is describing as the fallen ones, dark spirits, angels who willfully chose not to keep their estate with our Creator, stars in heaven that were extinguished, ships that became wrecked due to their own wicked and perverse ways, trees burned with fire, these are all WAYS OF SAYING that one third of His Creation by their own free will REFUSED/REFUSES to live righteously and peacefully with one another and instead chooses to harm one another and our Creator is then clear that those who do so wickedly without repentance must be SEPARATED FROM the rest of His Creation, CONFINED and DESTROYED. That is what all the separation between light and darkness, clean and unclean, etc. is about; spiritually speaking, BUT HE MAKES HIS PHYSICAL UNIVERSE DEMONSTRATE spiritual TRUTH so that we can learn from Him not just by reading the Holy Bible but by making accurate observations (science and applying the scientific method) to His Creation!!!!!!!! When we listen to our Creator "trees in the Garden of Eden" are metaphors for living conscious beings that impart to us wisdom, knowledge and understanding. http://biblehub.com/ezekiel/31-18.htm and http://biblehub.com/proverbs/11-30.htm

When GOD, the TREE OF LIFE, is within a person, they think His Thoughts, Speak His Words, Do His Deeds, and so they become a tree planted in His Garden, a tree of life, watered by the Wisdom of His Holy Spirit. http://biblehub.com/jeremiah/17-8.htm and http://biblehub.com/psalms/1-3.htm

So what the Scriptures Reveal to Us All was that it was NECESSARY for our CREATOR to INCARNATE and DIE http://biblehub.com/john/12-24.htm here on earth; so that HE could PLANT HIMSELF http://biblehub.com/john/15-5.htm, the ONE TRUE TREE OF LIFE , and take ROOT http://biblehub.com/revelation/22-16.htm in the earth and GROW such that the whole world could partake! As we come to the TREE of LIFE and partake of His Words of Life, we UNITE with GOD! http://biblehub.com/john/17-21.htm and so becoming branches on the TREE OF LIFE the words of GOD that flow from us in Truth HEAL the nations!

http://biblehub.com/revelation/22- 2.htm and http://biblehub.com/john/15-5.htm

(Divine Revelations (the Book of Revelation) are about REALITY, images that REPRESENT OBSERVABLE TRUTH; BUT by stating this truth, I am NOT discounting Paradise, trees in Paradise, rivers in Paradise, etc. I am stating GOD's WORD is ALL encompassing about spiritual and physical things, general pervasive and detailed specific things ALL AT ONCE simultaneously expressed; covering ALL aspects of HIS Creation from individual cells, individual souls, to towns, cities, nations, the world and even the whole universe!) And by ABIDING IN OUR CREATOR AND HE IN US, others partake of this WATER OF LIFE that FLOWS from HE WHO INHABITS ETERNITY and HEAVEN ABOVE, yet REVEALED HIMSELF in EARTH BELOW, among the waters, the sea of Humanity, the CREATOR WALKING AND TALKING IN THE MIDST OF HIS GARDEN, HIS CREATION!

http://biblehub.com/john/1- 10.htm and

https://www.blastthetrumpet.org/PublicLetters/AAAUpdatedPublicAlertsMattersofLifeandDeath/Updates053016/GOD%20Part%201.pdf - God part 1

https://www.blastthetrumpet.org/PublicLetters/AAAUpdatedPublicAlertsMattersofLifeandDeath/Updates053016/GOD%20Part%202.pdf - God part 2

https://www.blastthetrumpet.org/PublicLetters/AAAUpdatedPublicAlertsMattersofLifeandDeath/Updates053016/GOD%20IS%20GOOD%20part%201%20of%202.pdf - God is Good part 1

https://www.blastthetrumpet.org/PublicLetters/AAAUpdatedPublicAlertsMattersofLifeandDeath/Updates053016/GOD%20IS%20GOOD%20part%202%20of%202.pdf - God is Good part 2

So just as GOD, our CREATOR, is the TREE of LIFE, the VINE of LIFE, the ROOT of LIFE, the SOURCE of TRUTH, WISDOM, KNOWLEDGE, UNDESTANDING AND ALL VIRTUES, the first to conceive of doing wickedly, satanil, is the tree of the knowledge of good and evil. (devils know the difference between righteousness and wickedness, but they willfully choose to do wickedly) GOD chooses to declare only TRUTH, but lies and deceptions came from the first to imagine or conceive of them, satanil. When Eve or ANYONE listens to any consciousness, spirit, that tells them a lie or deception, and acts upon that lie or deception, they are eating from the tree of the knowledge of evil, they are choosing to partake of and be like the devil! ignorant, deceptive, wicked, rebellious, arrogant, murderous, etc. THIS IS WHY OUR CREATOR TOLD US PLAINLY WE COULD DISTINGUISH BETWEEN THE TRUE AND THE FALSE BY WHAT THEY DO! (their fruits, the results of listening to them are SHOWN plainly for all to observe) -

https://www.blastthetrumpet.org/PublicLetters/AAAUpdatedPublicAlertsMattersofLifeandDeath/Updates053016/BibleVquranTruthVlies.pdf - The HOLY BIBLE v. the quran and

Matthew 22:36-40

36 "Teacher, which is the great commandment in the Law?" 37 And he said to him, "You shall love the Lord your God with all your heart and with all your soul and with all your mind. 38 This is the great and first commandment. 39 And a second is like it: You shall love your neighbor as yourself. 40 On these two commandments depend all the Law and the Prophets."

God commands everyone to Love Him and EACH OTHER, so nothing in the Holy Bible encourages evil of any kind! Whereas the quran:

https://www.thereligionofpeace.com/pages/quran/index.aspx - MUCH evil! so, the quran is not from God but from the devil.

https://www.blastthetrumpet.org/PublicLetters/AAAUpdatedPublicAlertsMattersofLifeandDeath/Updates053016/Refutations%20of%20the%20False%20Accusations%20Made%20By%20Worldviews%20Against%20Christ%20and%20His%20Followers.pdf - answering common objections and false accusation against Christianity and

https://www.blastthetrumpet.org/PublicLetters/AAAUpdatedPublicAlertsMattersofLifeandDeath/Updates053016/satanil%20v%20GOD%20part%202.pdf - how the devil inspires those false accusations against Christ and His Followers, Christians.

I mention all these things because I want people to understand our Creator is revealing FAR MORE to us in the Creation account than just physical observations. I also want to interject here about how many scoffers and mockers these days laugh at the notion of Intelligent Design or claim if that if design is true, then it is not so clever and begin to point at all the many things in our present-day existence that are wicked, evil, or corrupted in nature. WHAT I HAVE NEVER HEARD ANY CREATIONIST TELL THESE MOCKERS IS THAT PRESENT-DAY CREATION IS WHAT WE OBSERVE AS A RESULT OF THE FALL! Death, decay, even our bodies growing old and frail is all designed as the intentional consequences of departing from the TREE of LIFE; in order to partake of the tree of the knowledge of evil!

(the whole world lies in darkness - http://biblehub.com/ephesians/2- 2.htm and http://biblehub.com/1_john/5-19.htm and ALL have sinned - http://biblehub.com/romans/3- 23.htm)

And the earth became cursed! (suffering and hard labor and many other consequences, poisonous plants, unclean animals, thorns and weeds, briars and awful things entered present day earth such that the DESIGN PRESENTLY OBSERVED is NOT THE INTENDED FINISHING WORK OF GOD, but reflects Blessings and cursings, Good and evil, Life and death in all the world)! All their usual examples of "poor design" that I have ever heard any of them state are the direct results of disobedience, sins, crimes and evil deeds, and some even mistakenly think

GOD is to blame for those who are PURPOSEFULLY corrupting His Creation through genetically modified organisms!

https://www.blastthetrumpet.org/PublicLetters/AAAUpdatedPublicAlertsMattersofLifeandDeath/Updates053016/Why%20So%20Much%20Suffering.pdf and

https://www.blastthetrumpet.org/PublicLetters/AAAUpdatedPublicAlertsMattersofLifeandDeath/Updates053016/Crimes%20Against%20Humanity%20Part%201.pdf - Crimes against humanity - genetic modification of Divine Designs.

and
https://www.blastthetrumpet.org/PublicLetters/AAAUpdatedPublicAlertsMattersofLifeandDeath/Updates053016/The%20Ten%20Commandments%20Part%201.pdf - Ten Commandments Part 1

https://www.blastthetrumpet.org/PublicLetters/AAAUpdatedPublicAlertsMattersofLifeandDeath/Updates053016/The%20Ten%20Commandments%20Part%202.pdf - Ten Commandments Part 2 https://www.blastthetrumpet.org/PublicLetters/AAAUpdatedPublicAlertsMattersofLifeandDeath/Updates053016/The%20Ten%20Commandments%20Part%203.pdf - Ten Commandments Part 3

As such, presently, mankind is given TEMPORAL bodies INTENTIONALLY (these earthly tabernacles in which our spirits dwell); such that their days are numbered on earth PURPOSEFULLY; so that those CHOOSING TO DO WICKEDLY are unable to do so perpetually; while the RIGHTEOUS will be given the PROPER and INTENDED DESIGNS of our CREATOR of IMMORTAL and GLORIFIED BODIES that defy even what we can currently imagine!

https://www.blastthetrumpet.org/PublicLetters/AAAUpdatedPublicAlertsMattersofLifeandDeath/Updates053016/The%20Difference%20Between%20Spirit%20and%20Glorified%20Body.pdf - Our Glorified Bodies is one of the last things God does in perfecting us!

So don't pay attention to mockers, who can't even make one protein of a living cell from known components, let alone the entire universe and yet still have the audacity to critique our Creator. So, the entire first chapter of Genesis, the Biblical Creation Account is describing Creative Thought concerning EVERYTHING that GOD intended to MAKE MANIFEST. And tells how in the void of space, mass, matter, energy, living things were all brought into existence over what we think of as "time". So the seven days are related more to categories of creation (light, darkness, water, earth, space, void, energy, mass, matter formed, plants, animals, mankind, work, rest, days, nights, evenings, mornings, time, stars, heaven, signs in the heavens, signs in the earth, etc.), CREATIVE THOUGHT BEING MADE MANIFEST OVER TIME, than all things that happened all at once thousands of years ago or longer. Every newborn male or female child and all newborn land animals are part of the sixth day, every tree, plant and herb that springs from the ground is part of the third day, every newborn fish or bird into our midst part of the fifth day, and so on each day has categories of aspects about all Creation and so the days or generations or hosts of heaven and earth are all fulfilled OVER THE PASSAGE OF TIME! Likewise, the Blessings and Cursings, the Judgments, in the Book of Revelation that over time ALL THOSE THINGS COME TO PASS. I meet futurists constantly who, even though the Book of Revelation was written almost 2000 years ago, STILL think it is ONLY talking about future events; when it is VERY CLEAR that the Revelations are about things past, present and future FROM THE MOMENT they were shown the Revelator and recorded as such!

http://biblehub.com/revelation/1-19.htm and http://biblehub.com/revelation/17-10.htm clearly the topics in the Book are Revelations about ALL OF HISTORY concerning ALL OF CREATION. Likewise, Genesis is also about ALL of CREATION; not just events long ago. In fact, the Words of our Creator permeate His Creation (reveal to us "reality" concerning ALL things, past, present and future).

http://biblehub.com/hebrews/1-3.htm and

http://biblehub.com/1_corinthians/2-10.htm and

http://biblehub.com/hebrews/11-3.htm:

http://biblehub.com/hebrews/1-3.htm The Supremacy of the Son ...2in these last days has spoken to us in His Son, whom He appointed heir of all things, through whom also He made the world.3And He is the radiance of His glory and the exact representation of His nature, and upholds all things by the word of His power. When He had made purification of sins, He sat down at the right hand of the Majesty on high, 4having become as much better than the angels, as He has inherited a more excellent name than they....

http://biblehub.com/john/1-1.htm The Beginning 1In the beginning was the Word, and the Word was with God, and the Word was God. 2He was in the beginning with God.

3All things came into being through Him, and apart from Him nothing came into being that has come into being....

But getting back to the Third Day and what we observe. http://biblehub.com/genesis/1-9.htm

Take a look at properties of water in the space station: https://www.youtube.com/watch?v=KFPvdNbftOY

and the bubble inside the honey https://www.youtube.com/watch?v=2vio09T-8qA

and how the water forms spheres when not attached to solids https://www.youtube.com/watch?v=RmV90BmMNMg So

liquids in space (superheated elements in a liquid state) forms spheres;

https://www.youtube.com/watch?v=LIFc351sCzI

(stars are lakes of fire in space or stones of fire in space).

https://www.youtube.com/watch?v=ntQ7qGilqZE and https://www.youtube.com/watch?v=DKxAs_f1SP8

With myriads of such lakes of fire, it is ridiculous to doubt the existence of the Lake of Fire in the Biblical Account; in the Book of Revelation.
http://biblehub.com/revelation/21-8.htm

Also note that water adheres to solids and forms or flows over odd shaped structures (so water on earth apart from other forces or forms of energy, should cover the entire surface of our planet). We know therefore that the mass or matter of earth that is "solid" is therefore spherical in shape and that, MORE THAN LIKELY, suggests the entire earth was at one time in a liquid state in space. (probably superheated matter that has cooled over time) But my point in showing the space station experiments in relation to the separation of water and dry land; is to show that there are forces governing our earth that are more than just physical properties. More than just dark matter, dark energy, gravity, molecular adhesion, etc. But that our CREATOR is CONSTANTLY PRESENT defining boundaries.

https://www.biblegateway.com/passage/?dc=2&utm_expid=13466113-14.7HxyD5XxTMyyfz37nhlGg.2&search=Job+38&utm_referrer=https%3A%2F%2Fwww.google.com%2F Job 38 Amplified Bible (AMP) 38 Then the Lord answered Job out of the whirlwind and said, 2 Who is this that darkens

counsel by words without knowledge? 3 Gird up now your
loins like a man, and I will demand of you, and you declare
to Me. 4 Where were you when I laid the foundation of
the earth? Declare to Me, if you have and know
understanding. 5 Who determined the measures of the
earth, if you know? Or who stretched the measuring line
upon it? 6 Upon what were the foundations of it fastened,
or who laid its cornerstone, 7 When the morning stars
sang together and all the sons of God shouted for joy? 8
Or who shut up the sea with doors when it broke forth and
issued out of the womb?— 9 When I made the clouds the
garment of it, and thick darkness a swaddling band for it,
10 And marked for it My appointed boundary and set bars
and doors, 11 And said, Thus far shall you come and no
farther; and here shall your proud waves be stayed? 12
Have you commanded the morning since your days began
and caused the dawn to know its place, 13 So that [light]
may get hold of the corners of the earth and shake the
wickedness [of night] out of it? 14 It is changed like clay
into which a seal is pressed; and things stand out like a
many-colored garment. 15 From the wicked their light is
withheld, and their uplifted arm is broken. 16 Have you
explored the springs of the sea? Or have you walked in the
recesses of the deep? 17 Have the gates of death been
revealed to you? Or have you seen the doors of deep
darkness? 18 Have you comprehended the breadth of the
earth? Tell Me, if you know it all. 19 Where is the [a]way
where light dwells? And as for darkness, where is its
abode, 20 That you may conduct it to its home, and may
know the paths to its house? 21 You must know, since you

were born then! Or because you are so extremely old! 22 Have you entered the treasuries of the snow, or have you seen the treasuries of the hail, 23 Which I have reserved for the time of trouble, for the day of battle and war? 24 By what way is the light distributed, or the east wind spread over the earth? 25 Who has prepared a channel for the torrents of rain, or a path for the thunderbolt, 26 To cause it to rain on the uninhabited land [and] on the desert where no man lives, 27 To satisfy the waste and desolate ground and to cause the tender grass to spring forth? 28 Has the rain a father? Or who has begotten the drops of dew? 29 Out of whose womb came the ice? And the hoary frost of heaven, who has given it birth? 30 The waters are congealed like stone, and the face of the deep is frozen. 31 Can you bind the chains of [the cluster of stars called] Pleiades, or loose the cords of [the constellation] Orion? 32 Can you lead forth the signs of the zodiac in their season? Or can you guide [the stars of] the Bear with her young? 33 Do you know the ordinances of the heavens? Can you establish their rule upon the earth? 34 Can you lift up your voice to the clouds, so that an abundance of waters may cover you? 35 Can you send lightnings, that they may go and say to you, Here we are? 36 Who has put wisdom in the inward parts [or in the dark clouds]? Or who has given understanding to the mind [or to the meteor]? 37 Who can number the clouds by wisdom? Or who can pour out the [water] bottles of the heavens 38 When [heat has caused] the dust to run into a mass and the clods to cleave fast together? 39 Can you [Job] hunt the prey for the lion? Or satisfy the appetite of the young lions 40 When they couch in their dens or lie in wait in their hiding place? 41 Who provides for the raven its prey when its young ones cry to God and wander about for lack of food?

https://www.youtube.com/watch?v=v8JflYUt89w

Scientists may CLAIM all these things about observable reality are "natural"; but I argue back that NOTHING we observe is such! REALITY IS NOT JUST SUPERNATURALLY INITIATED BUT SUPERNATURALLY SUSTAINED AND PRESENTED TO US ALL MOMENT BY MOMENT! That it is GOD our CREATOR who gives us all these so called "natural" properties, forms of energy, elements, mass, matter, and causes the very formations of not only everything we observe in heaven above but on earth below MOMENT BY DIVINE MOMENT!

https://www.youtube.com/watch?v=PwPI-yCxOwA&list=UUbl84DzN_SB2IWLLA10XRrQ

It is GOD in all through all, above all, surrounding all, that is EVER PRESENT as the CAUSE for ALL that we observe! It takes multiple factors to keep the water in the oceans and seas from doing just like we saw on the space station and covering the mountains and valleys of the surface of the earth! Despite everything I just pointed out about the balance or fine tuning that keeps stars, planets and even atoms from collapsing, it is likely you will still find people snickering "like gravity" to themselves; even though not one of them truly understands "gravity" or for that matter the reasons WHY for all the various properties of all the various forms of energy and matter; because that knowledge comes by KNOWING our CREATOR. We can observe that evolutionists, like Dawkins, are content with NOT KNOWING and NOT FINDING AND ASCERTAINING FUNDAMENTAL TRUTH in favor of their delusionary theories in the name of science; but CREATIONISTS were NOT satisfied with rubbish postulated in the Name of Science; especially in the fields of biology and modern theoretical physics and so Biblical Creationists went BEYOND being content with only observing "natural" phenomenon; in search for the reasons WHY various forms of energy and mass have certain properties; and when they went searching BEYOND "what we EASILY observe" in thorough, diligent quest for the CAUSE, they did not IMAGINE or merely choose to believe "well GOD did it"; but DISCOVERED A PLETHORA OF EMPIRICAL, VALID EVIDENCE that PROVES HE DID! -

https://www.blastthetrumpet.org/PublicLetters/AAAUpdatedPublicAlertsMattersofLifeandDeath/Updates053016/The%20Dawkins%20Delusion.pdf - Dawkins wrote a book called the God delusion it was reflective upon His own God-denying conscience.

I reiterate! Please help people not to present themselves as

https://www.blastthetrumpet.org/PublicLetters/AAAUpdatedPublicAlertsMattersofLifeandDeath/Updates053016/Mentally%20Challenged.pdf by teaching the Truth!

https://www.blastthetrumpet.org/PublicLetters/AAAUpdatedPublicAlertsMattersofLifeandDeath/Updates053016/Evidence%20of%20GOD.pdf and

https://www.blastthetrumpet.org/PublicLetters/AAAUpdatedPublicAlertsMattersofLifeandDeath/Updates053016/The%20Sufferings%20of%20Christ.pdf

It takes INTELLIGENCE and POWER that has defined and gives us the various distinguishable properties of all the forms of energy and mass; for very specific purposes!

http://www.godandscience.org/apologetics/designun.html and

http://www.reasons.org/articles/fine-tuning-for-life-in-the-universe

One would surmise that with the advent of weather warfare and geo-engineering that everyone on earth would realize that Truth by now!

https://www.blastthetrumpet.org/PublicLetters/AAAUpdatedPublicAlertsMattersofLifeandDeath/Updates053016/Crimes%20Against%20Humanity%20Part%202.pdf - crimes against humanity, geoengineering with chemicals harmful to us and life on earth

http://www.indiana.edu/~geog109/topics/10_Forces&Winds/GasPressWeb/PressGasLaws.html - notice that a rise in pressure causes a rise in temperature (it is not absolutely necessary to have an external heat source to make coal or oil, just sufficient pressure). "The Ideal Gas Law or Equation of State; The example used to illustrate Charles's law probably does not follow Charles's law exactly. It is very likely that, during the heating process, when the rate of collisions by the gas molecules increased, the pressure increased as well as the volume. Thus, in practical situations all three variables involved in Boyle's and Charles's law are linked and both principles are in action at the same time: Pressure: P Temperature: T Density: r = m/V These variables describe the state of the gas at any one time and are combined in the single relationship known as the ideal gas law or the equation of state: or where R, the constant of proportionality, is known as the universal gas constant (= 287 J·kg-1·K-1) When mass and pressure are held constant, the gas law collapses to Charles's law. When mass and temperature are held constant, it is equivalent to Boyle's law. Thus, the gas law combines the two laws. The equation of state or ideal gas law is one of the most fundamental relationships linking the three variables temperature, pressure and density that describe the thermodynamic state of the atmosphere. Some application of the gas law is involved in most practical problems in meteorology." -

http://www.indiana.edu/~geog109/topics/10_Forces&Winds/GasPressWeb/PressGasLaws.html As anyone can read pressure is directly proportional to temperature; a dramatic rise in pressure equals a dramatic rise in temperature. So massive rapid deposits of mud over animals and vegetation IS A PLAUSIBLE EXPLANATION for coal and oil beds THAT CAN FORM RAPIDLY under such intense pressure! Hence, we find artifacts in coal deposits that, according to evolutionists, are supposed to be hundreds of millions of years old.

https://www.google.com/search?q=bell+and+other+artifacts+found+in+coal+deposits&rlz=1C1GIGM_enUS535US535&espv=2&biw=1066&bih=702&tbm=isch&tb

o=u&source=univ&sa=X&ei=lvpbVJmABsOnyATP_ID4Cg&ved=0CB0QsAQ

http://www.discoverynews.us/DISCOVERY%20MUSEUM/AncientMan/The_Missing_Links_Display3.html

https://www.google.com/webhp?sourceid=chrome-instant&ion=1&espv=2&ie=UTF8#q=man%20made%20artifacts%20in%20coal precious gems

http://nature.berkeley.edu/classes/eps2/wisc/Lect3.html#metam can all be created rapidly in labs:

http://www.gia.edu/gem-synthetic and

https://www.google.com/webhp?sourceid=chromeinstant&rlz=1C1GIGM_enUS535US535&ion=1&espv=2&ie=UTF8#q=lab%20gems%20can%20be%20created%20rapidly

So, we have two points here that I want to make. 1) When man has PROVEN all "naturally occurring" substances can be created rapidly, IT IS UNREASONABLE then to think that deep time is mandated for creation as it is observed! 2) that here again we have evidence that

WHAT WE OBSERVE has BEEN DESIGNED to TEACH US SPIRITUAL TRUTH in that by heat and pressure common substances are transformed into precious gems. GOD puts it this way in His Word to us:

http://biblehub.com/zechariah/9-16.htm And the LORD their God will save them in that day as the flock of His people; For they are as the stones of a crown, Sparkling in His land.

http://biblehub.com/1_peter/4-12.htm

1 Peter 4:12-19Authorized (King James) Version (AKJV) 12 Beloved, think it not strange concerning the fiery trial which is to try you, as though some strange thing happened unto you: 13 but rejoice, inasmuch as ye are partakers of Christ's sufferings; that, when his glory shall be revealed, ye may be glad also with exceeding joy. 14 If ye be reproached for the name of Christ, happy are ye; for the spirit of glory and of God resteth upon you: on their part he is evil spoken of, but on your part he is glorified. 15 But let none of you suffer as a murderer, or as a thief, or as an evildoer, or as a busybody in other men's matters. 16 Yet if any man suffer as a Christian, let him not be ashamed; but let him glorify God on this behalf. 17 For the time is come that judgment must begin at the house of God: and if it first begin at us, what shall the end be of them that obey not the gospel of God? 18 And if the

righteous scarcely be saved, where shall the ungodly and the sinner appear? 19 Wherefore let them that suffer according to the will of God commit the keeping of their souls to him in well doing, as unto a faithful Creator.

http://biblehub.com/malachi/3-3.htm The Coming Day of Judgment ...2"But who can endure the day of His coming? And who can stand when He appears? For He is like a refiner's fire and like fullers' soap.3"He will sit as a smelter and purifier of silver, and He will purify the sons of Levi and refine them like gold and silver, so that they may present to the LORD offerings in righteousness. 4"Then the offering of Judah and Jerusalem will be pleasing to the LORD as in the days of old and as in former years....

The LORD is not only TELLING US, but SHOWS US IN HIS CREATION, that the pressures that comes upon us in life, the tough decisions, the trials and temptations ALL SERVE to purify us, to perfect us, to transform us from what is common or even what is discarded or burned as rubbish into precious creations of inestimable value!

http://biblehub.com/james/1-3.htm Rejoicing in Trials 2Consider it all joy, my brethren, when you encounter various trials,3knowing that the testing of your faith produces endurance. 4And let endurance have its perfect result, so that you may be perfect and complete, lacking in nothing. 5But if any of you lacks wisdom, let him ask of God, who gives to all generously and without reproach, and it will be given to him....

http://biblehub.com/romans/5-4.htm Faith Brings Peace and Joy ...3And not only this, but we also exult in our tribulations, knowing that tribulation brings about perseverance; 4and perseverance, proven character; and proven character, hope; 5and hope does not disappoint, because the love of God has been poured out within our hearts through the Holy Spirit who was given to us.

Every time we are put to the test, GOD is purging away dross, (exposing and removing sin from our lives), He is transforming our heart and mind, our overall character to become more and more Virtuous, Honorable, Radiant Beings that Rejoice Before Him now and forever!!!!!!!! ALL the observable universe TEACHES US ABOUT THE CHARACTER OF OUR CREATOR, TEACHES us HOW to discern right from wrong, good from evil, not just in HIS WORD, the HOLY BIBLE, but EVERYTHING WE OBSERVE!!! THE OVERWHELMING EVIDENCE FOR GOD AND WHO HE IS, IS BEFORE ALL OF US UBIQUITOUSLY TO PERCEIVE 1) HIS ENTIRE CREATION TESTIFIES ABOUT HIM 2) HIS WORD TESTIFIES ABOUT HIM 3) HIS MESSENGERS TESTIFY ABOUT HIM 4) GOD TESTIFIES ABOUT HIMSELF!!!! I've heard people try to disregard verbal and written testimony as scientific evidence or even legal evidence, but I find such claims LUDICROUS attempts to WILLFULLY REMAIN IGNORANT!

https://www.facebook.com/photo.php?fbid=861703013874470&set=gm.74379098570465 0&type=1&theater

KNOWLEDGE is COMMUNICATED and therefore TESTIMONY; ESPECIALLY TESTIMONY MAINTAINED IN THE PRESENCE OF EXTREME SUFFERING, TORTURE AND EVEN DEATH is ABSOLUTELY "scientific" evidence. (The earliest Apostolic Martyrs were put to death for maintaining their testimony of WHAT THEY PERSONALLY OBSERVED; even if the millions of martyrs that have followed might have suffered for what they believed. I say, "might" because from the perspective of the unbelieving world they cannot "see" the Spirit of GOD in His People down through the ages, and they cannot know whether or not the many martyrs over the centuries also saw CHRIST or not; even though many who suffered and died stated that was so.) We are not talking about one or two isolated events in history but MASS MILLIONS!

http://www.telegraph.co.uk/science/science-news/11144442/First-hintof-life-after-death-in-biggest-ever-scientific-study.html

Beyond "death" and back testimonies, and testimonies of who the LIVING GOD is even under torture and death! http://homecomers.org/mirror/ MILLIONS SUFFERED AND DIED RATHER THAN RECANT ONE TINY TRUTH ABOUT GOD! And it is STILL happening all over the world!

https://www.blastthetrumpet.org/PublicLetters/AAAUpdatedPublicAlertsMattersofLifeandDeath/Updates053016/islam%20is%20Evil%20part1.pdf - Whereas Christianity is built upon Loving God and Loving Each Other (Christ's Two Greatest Commandments), not all religions teach that. islam for example teaches muslims to commit crimes in its doctrines and practices. https://www.thereligionofpeace.com/pages/quran/index.aspx - Christ said that which lies and teaches murder is of the devil.

All those who would CLAIM such testimony is not valid evidence, should ask themselves truthfully, if they would make such a claim under the same penalties of torture and death that those who testified of CHRIST have endured for His Name's sake. (Is what you believe such sure and precious truthful KNOWLEDGE that you would endure such atrocities defending it; even when if you knew all you had to do was verbally deny that belief and the torture would stop immediately?) BUT all those valiant men and women SEALED THEIR TESTIMONIES BEFORE HEAVEN AND EARTH EXACTLY AS PROPHESIED!

http://biblehub.com/revelation/12-11.htm ...10Then I heard a loud voice in heaven, saying, "Now the salvation, and the power, and the kingdom of our God and the authority of His Christ have come, for the accuser of our brethren has been thrown down, he who accuses them before our God day and night. 11"And they overcame him because of the blood of the Lamb and because of the word of their testimony, and they did not love their life even when faced with death.

In fact, once a person is truly "born again", the TRUTH of our CREATOR is so blindingly obvious, that they all wonder HOW WAS IT EVEN POSSIBLE THEY COULD HAVE BEEN IN SUCH MASSIVE IGNORANCE PRIOR! (Until they read on about the deceiver and how they were all intentionally brainwashed to believe in utter nonsense that defies what can be observed ubiquitously - http://biblehub.com/2_corinthians/4-4.htm and https://www.blastthetrumpet.org/PublicLetters/AAAUpdatedPublicAlertsMattersofLifeandDeath/Updates053016/Brainwashing.pdf) But our ignorance is not just due to deceptions and delusions, but the fact that creation again demonstrates TRUTH in the process of birth and maturation. An egg in the womb does not even come to life until it is fertilized with the seed of a man; NOR until anyone receives CHRIST are they born of His Spirit; as such, it is IMPOSSIBLE for them to perceive the entire spiritual realm of the universe or even understand His Words! (just like an egg in a womb has no consciousness about the physical world until it is fertilized and then born into this world and then matures to explore and perceive it) Even so, those "born again" over time mature to understand more and more about TRUTH, GOD our CREATOR and His Creation, until they are born into Paradise itself, or until all Creation manifests God's Sons and Daughters. Rom 8:21-23 your birth into your New Creation/Paradise begins with the death/dying of this old, sinful, fallen body/creation. Eph 2:1-10, Jn 3:1-15, 1 Cor 15:31, 1 Cor 9:27, Rom 7:24,25

https://www.blastthetrumpet.org/PublicLetters/AAAUpdatedPublicAlertsMattersofLifeandDeath/Updates053016/You%20MUST%20Be%20Born%20Again.pdf YOU MUST BE BORN AGAIN! and https://www.blastthetrumpet.org/PublicLetters/AAAUpdatedPublicAlertsMattersofLifeandDeath/Updates053016/KNOW%20THE%20LORD%20JESUS%20CHRIST.pdf - KNOW the LORD JESUS CHRIST!

http://www.evolutionnews.org/2012/05/has_the_talk-or059171.html

As a Biblical Creationist, I have had unenlightened persons mock me for referencing talkorigins in my notes or even presentations like zeitgeist or other opponents to the historicity and veracity of the Holy Bible. If Biblical Creationists try and pretend arguments (however ridiculous) against the Divine Authority of the Holy Bible do not exist, we only then appear as much in denial of reality as the opponents (God-haters) themselves. atheists and evolutionists need to understand, Biblical Creationists not only address their fiction but soundly refute it; even their most cherished cultic apologetics.

http://www.anncoulter.com/columns/2011-08-31.html

Some post nonsense, like quoting dawkins and his cronies claiming they find transitional forms in predicted geologic layers, when that simply is not the case. Complex life forms are found across multiple layers and appear suddenly within the layers of presumed deep time. There is no fictional first came reptiles, then amphibians, etc. like evo's constantly imagine in their drawings and animations so when they say they go to such and such a layer and find amphibians that is like DUH! amphibians and reptiles still coexist today. The fossil record is no friend to dawkins or darwinian nonsense at all.

http://www.darwinthenandnow.com/2013/05/richard-dawkins-dumps-the-fossil-record/

Understand the genetics involved in defining each of the various life forms. Information that coordinates in the case of a human being TRILLIONS of cells all working together to make the form of a man as opposed to a rhino, fish, ape, etc. "transitioning" between such forms would require not one mutation but mutations of epic proportions all coordinated to make an actual functional change (as opposed to what has been observed that mutations/defects in DNA/RNA are HARMFUL and tend to lead to the premature demise of the affected organism; not its propagation).

https://www.youtube.com/watch?v=WOYpByOLWwo that is the reason for the silence of those acolytes of the cult of evolution.

https://www.blastthetrumpet.org/PublicLetters/AAAUpdatedPublicAlertsMattersofLifeandDeath/Updates053016/Evolution%20and%20Atheism%20Intertwined%20Cults%20of%20the%20Insane.pdf - evolution and atheism causes people not to present themselves as having a sound mind, but instead are in denial of reality.

Evolutionists would have people believe in impossibilities; in a desperate attempt to avoid reality (GOD, our CREATOR). And as such, they want to IMAGINE there are geologic layers of deep time in which first one kind of organism appeared and then another and so on; but I have just showed numerous examples that is all IMAGINATION and has NOTHING TO DO WITH WHAT WE OBSERVE. Fossils of plants, animals, single celled organisms and even human artifacts all are buried in the SAME LAYERS and even ACROSS MULTIPLE LAYERS!

https://www.youtube.com/watch?v=8RYgWoLScng

https://www.google.com/webhp?sourceid=chromeinstant&rlz=1C1GIGM_enUS535US535&ion=1&espv=2&ie=UTF8#q=fossils%20of%20modern%20birds%20and%20dinosaurs%20found%20together

https://www.google.com/webhp?sourceid=chromeinstant&rlz=1C1GIGM_enUS535US535&ion=1&espv=2&ie=UTF8#q=human%20artifacts%20found%20in%20fossil%20beds

https://www.google.com/webhp?sourceid=chromeinstant&rlz=1C1GIGM_enUS535US535&ion=1&espv=2&ie=UTF-8#q=human+sandal+print+with+trilobite

https://www.google.com/webhp?sourceid=chromeinstant&rlz=1C1GIGM_enUS535US535&ion=1&espv=2&ie=UTF8#q=soft+tissue+dinosaur+fossils+evidence+hardens

https://www.google.com/webhp?sourceid=chromeinstant&rlz=1C1GIGM_enUS535US535&ion=1&espv=2&ie=UTF8#q=list%20of%20soft%20tissue%20dinosaurs%20found

http://www.dailymail.co.uk/news/article-2034007/Worlds-largest-crocodile-captured-Philippinesvillagers.html - if these creatures were extinct, there is no doubt they would be called a dinosaur.

https://www.google.com/search?q=mammal+tracks+found+together+with+the+tracks+of+dinosaurs&rlz=1C1GIGM_enUS535US535&oq=mammal+tracks+found+together+with+the+tracks+of+dinosaurs&aqs=chrome..69i57&sourceid=chrome&es_sm=93&ie=UTF-8

http://creation.com/ichthyosaur-birth-graveyard-soft-tissue

https://www.youtube.com/watch?v=R1GJjHXX0Ic

https://www.youtube.com/watch?v=dWe3cteDuBc

https://www.facebook.com/photo.php?fbid=513395842152410&set=gm.939309672819446&type=3&t heater

https://www.youtube.com/watch?v=y9WImEIE_hk

http://newgeology.us/presentation48.html - IT IS NOT SCIENCE OR SCIENTIFIC TO DISMISS DATA AND EVIDENCE JUST BECAUSE IT DOESN'T CONFIRM YOUR CHOSEN DELUSIONS! (the cult of evolution and atheism has been intentionally suppressing scientific evidence supporting the Biblical Creation account).

https://www.youtube.com/watch?v=QbNpi2d5oEs

https://www.youtube.com/watch?v=niDCq3TbvOo

https://www.youtube.com/watch?v=aESGdORhHlg&feature=share - NOTE AND REMEMBER THE FOSSIL RECORD SUPPORTS BIBLICAL CREATION ACCOUNT AND THE GLOBAL FLOOD HISTORIC FACT; it in no way supports the fantasies of evolutionists. Every one needs to watch Biblical Creation scientists present evidence to the world that the Holy Bible is ACCURATE, the flood took place! https://www.youtube.com/watch?v=CcgXKHUbWB4 – the same scientists that argue for the historicity of the Holy Bible, put forth a whole series on the subject, explaining why the bent layers in the Grand Canyon and other strata all over the world are best explained by the recorded history in the Holy Bible of a global flood, not that those sedimentary layers are hundreds of millions to billions of years old or apart. (That is just lying nonsense put forth by God-deniers and mockers of the what the Gospel of JESUS CHRIST, the Holy Bible, PLAINLY TELLS US ALL!)

https://youtu.be/niDCq3TbvOo

Actually, there is considerable evidence that man and dinosaurs/dragons coexisted as previously cited and in review:

https://www.google.com/search?q=paluxy+tracks+answers+in+genesis&rlz=1C1GIGM_enUS535US535&oq=paluxy+trac&aqs=chrome.2.0j69i57j0l4.9425j1j8&sourceid=chrome&es_sm=93&ie=UTF8#q=paluxy+tracks+answers+in+genesis&tbm=vid

Again, evolutionists have been busted time and again for their intentional frauds and fictional claims in the name of science.

https://www.google.com/search?q=evolutionary+frauds&rlz=1C1GIGM_enUS535US535&oq=evolutionary+frauds&aqs=chrome..69i57j0l4.4711j1j8&sourceid=chrome&es_sm=93&ie=UTF-8

So, they also LIE and CLAIM that is what Biblical Creationists are doing rather than showing empirical evidence that we are; evolutionists work to deceive the masses not only by their lies and frauds but by suppression of genuine empirical evidence proving the Biblical Account is true.

https://www.google.com/search?q=evidence+dinosaurs+and+man+coexisted&rlz=1C1GIGM_enUS535US535&oq=evidence+dinosaurs+and+man+&aqs=chrome.0.0j69i57j0l3.7799j1j8&sourceid=chrome&es_sm=93&ie=UTF-8

https://www.google.com/search?q=smithsonian+gatekeeping+knowledge+of+giants+and+other+facts+from+the+public&rlz=1C1GIGM_enUS535US535&oq=smithsonian+g&aqs=chrome.3.69i57j0l2j69i59j0l2.9593j1j8&sourceid=chrome&es_sm=93&ie=UTF-8

https://www.facebook.com/photo.php?fbid=10155770172335234&set=gm.615233428617492&type= 1&theater

https://www.facebook.com/photo.php?fbid=10156188297665234&set=gm.1626594734258499&type =3&theater

Soft Tissue Dinosaur Specimens

It is done all the time by the antichrists currently in power and mass brainwashing innocent children in coerced public indoctrination and reinforced by controlled media.

https://www.google.com/search?q=alan+watt+media+brainwashing+black+ops+programming&rlz=1C1GIGM_enUS535US535&oq=alan+watt+media+brainwashing+black+ops+programming&aqs=chrome..69i57.16711j1j8&sourceid=chrome&es_sm=93&ie=UTF-8

https://www.google.com/search?q=public+education+mass+brainwashing+by+antichrists&rlz=1C1GIGM_enUS535US535&oq=public+education+mass+brainwashing+by+antichrists&aqs=chrome..69i57j69i64.9639j1j8&sourceid=chrome&es_sm=93&ie=UTF-8

UNDERSTAND IT IS IMPOSSIBLE FOR carbon 14 and soft tissue to survive for "millions of years" SO SOFT TISSUE FINDS IRREFUTABLY PROVE that deep time assumptions regarding the fictional geologic column ARE FALSE!!!!!!!! (evolutionists are LYING to children and BRAINWASHING THEM with FICTION; in the Name of Science!)

http://www.sciencedirect.com/science/article/pii/S0065128113000020

https://kgov.com/dinosaur-soft-tissue-original-biological-material - soft tissues defy evolutionary dogmas of deep time claims.

https://www.google.com/webhp?sourceid=chromeinstant&rlz=1C1GIGM_enUS535US535&ion=1&espv=2&ie=UTF8#q=mammals%20and%20dinosaurs%20coexisting

https://www.google.com/webhp?sourceid=chrome-instant&ion=1&espv=2&ie=UTF8#q=list%20of%20dinosaur%20soft%20tissue%20specimens

http://www.icr.org/article/8822

The entire evolutionary paradigm from cosmic origin to macro-evolution of less complex creatures coming spontaneously into existence (LIFE FROM NON-LIFE IS IMPOSSIBLE; violates the law of biogenesis) and then morphing into more complex species is ALL

IMAGINARY FICTION; not even good science fiction!!!!!!!! (and as shown in numerous references throughout this thesis, all their deep time claims of circular reasoning between fossils and rocks are FALSE)

https://www.google.com/webhp?sourceid=chromeinstant&rlz=1C1GIGM_enUS535US535&ion=1&espv=2&ie=UTF8#q=polonium%20halos%20unrefuted%20evidence

https://www.google.com/webhp?sourceid=chromeinstant&rlz=1C1GIGM_enUS535US535&ion=1&espv=2&ie=UTF8#q=fictional%20geologic%20column%20lies%20in%20the%20textbooks - I find it astonishing that with all the evidence I am disclosing that people STILL are debating this issue. THE OBSERVABLE EVIDENCE PROVES THAT PLANTS, ANIMALS, BIRDS, FISH, MAN, and yes EVEN DINOSAURS, GIANTS, NEPHILIM, REPHAIM, all have been on planet earth TOGETHER! (Again, WHAT WE OBSERVE, SCIENCE, SUPPORTS THE BIBLICAL ACCOUNT of CREATION; and does so OVERWHELMINGLY)!

https://www.google.com/webhp?sourceid=chromeinstant&rlz=1C1GIGM_enUS535US535&ion=1&espv=2&ie=UTF8#q=evidence+that+man+and+dinosaurs+coexisted+is+overwhelming

https://www.google.com/webhp?sourceid=chromeinstant
&rlz=1C1GIGM_enUS535US535&ion=1&espv=2&ie=UTF8#
q=evidence+that+man+and+dinosaurs+coexisted+is+over
whelming&tbm=vid

https://www.youtube.com/watch?v=1zz8_MxcnzY&list=PL
EA099CE59F0A0322&index=2

https://www.youtube.com/watch?v=CEmnXV5Qs2Q&list=
PLDD19870F568B7818

https://www.google.com/webhp?sourceid=chromeinstant
&rlz=1C1GIGM_enUS535US535&ion=1&espv=2&ie=UTF8#
q=giant%20human%20skeletons%20in%20newspapers%2
0around%20the%20world

https://www.google.com/webhp?sourceid=chromeinstant
&rlz=1C1GIGM_enUS535US535&ion=1&espv=2&ie=UTF8#
q=dinosaurs%20in%20newspapers%20alive%20in%20rece
nt%20history

http://worldnewsdailyreport.com/smithsonian-admits-to-destruction-of-thousands-of-giant-humanskeletons-in-early-1900s/ - satires are not necessarily false in the point they are making.

https://www.youtube.com/watch?v=CEmnXV5Qs2Q&list=PLDD19870F568B7818

https://www.youtube.com/watch?v=EEt59cBNsH0&list=PL6lzWC-yEtIfUNbshU1h9X4MHEu5QZWh4

Evidence of the Supernatural Presence of our CREATOR is in EVERYTHING; verifying the Biblical Account!

https://www.youtube.com/watch?v=ay_Db4RwZ_M

https://www.youtube.com/watch?v=rn73J9A0SnU

https://www.youtube.com/watch?v=LYBnJF2P_WQ

But if somehow that is not OVERWHELMINGLY CLEAR, let us continue to LOOK at REALITY; what WE OBSERVE, ACTUAL SCIENCE. The Third Day: Dry Ground ...11Then God said, "Let the earth sprout vegetation:

plants yielding seed, and fruit trees on the earth bearing fruit after their kind with seed in them"; and it was so.12The earth brought forth vegetation, plants yielding seed after their kind, and trees bearing fruit with seed in them, after their kind; and God saw that it was good. This is plainly observable truth. The genetic makeup of each and every organism, is passed on to reproduce its same kind of organism. Conifer trees by their seeds produce conifer trees, fruit trees by their seeds produce fruit trees, and even single celled organisms produce single celled organisms like the original. NEVER, NOT EVEN ONE TIME OUT OF BILLIONS AND BILLIONS OF OBSERVATIONS PRESENTLY AND THROUGHOUT HISTORY has one kind of organism produced another kind of organism. (like those who cling to the theory of evolution believe) Intentional genetic modification of anything is a crime of the highest order, an attempt to corrupt Divine Creation and such abominations bring about devastation ultimately.

https://www.blastthetrumpet.org/PublicLetters/AAAUpdatedPublicAlertsMattersofLifeandDeath/Updates053016/Crimes%20Against%20Humanity%20Part%201.pdf - crimes against humanity part 1

https://www.blastthetrumpet.org/PublicLetters/AAAUpdatedPublicAlertsMattersofLifeandDeath/Updates053016/Th

e%20Ten%20Commandments%20Part%201.pdf - The Ten Commandments part 1

https://www.google.com/webhp?sourceid=chromeinstant&rlz=1C1GIGM_enUS535US535&ion=1&espv=2&ie=UTF8#q=how%20gmos%20interrupt%20ecosystems%20and%20lead%20to%20extinctions

https://www.google.com/webhp?sourceid=chromeinstant&rlz=1C1GIGM_enUS535US535&ion=1&espv=2&ie=UTF-8#q=GMOs+causing+mass+die+offs

https://www.google.com/webhp?sourceid=chromeinstant&rlz=1C1GIGM_enUS535US535&ion=1&espv=2&ie=UTF8#q=gmos%20cause%20extinction%20events The abominations of genetic corruption that lead to the rise of the Nephilim and Rephaim on earth, perverted not just the thoughts and actions of creatures, but even their very essence. Such corruption was what brought on the global devastation of the flood. As such, genetic modification of ANYTHING, should be PERMANENTLY INTERNATIONALLY BANNED!!!!!!!! PRESERVE DIVINE DESIGNS! GOD DESIGNED ALL LIFE ON EARTH TO BRING FORTH LIKE KINDS FROM GENERATION TO GENERATION! Altering Divine Designs is like intentionally adding a virus into a functioning, intelligent designed computer, only GENECTIC

MODIFICATION of living organisms, is disrupting the ecology of earth! Ecosystems are suffering, and it only means that it is leading to our own demise!

http://biblehub.com/interlinear/genesis/6-11.htm

http://biblehub.com/interlinear/genesis/6-12.htm

http://biblehub.com/genesis/6-4.htm

https://www.youtube.com/watch?v=lktmmd7YnD8&index=1&list=PLEA099CE59F0A0322 - I address the flood in more detail in

https://www.blastthetrumpet.org/PublicLetters/AAAUpdatedPublicAlertsMattersofLifeandDeath/Updates053016/NO%20MORE%20BRAINWASHING%20p1.pdf No More Brainwashing, teach the Holy Bible, the Commands of our Eternal Creator, and

https://www.blastthetrumpet.org/PublicLetters/AAAUpdatedPublicAlertsMattersofLifeandDeath/Updates053016/Bill%20Nyes%20Science%20Lies.pdf I will touch on some more important facts within this thesis later on.

https://www.youtube.com/watch?v=_w-e_T1C9fM&list=PLDD19870F568B7818&index=8

https://www.facebook.com/1mill.creationist/photos/a.361018770595969.88428.286013238096523/875899082441266/?type=1&theater The Fourth Day: Sun, Moon, Stars 14Then God said, "Let there be lights in the expanse of the heavens to separate the day from the night, and let them be for signs and for seasons and for days and years; 15and let them be for lights in the expanse of the heavens to give light on the earth"; and it was so. 16God made the two great lights, the greater light to govern the day, and the lesser light to govern the night; He made the stars also.17God placed them in the expanse of the heavens to give light on the earth, 18and to govern the day and the night, and to separate the light from the darkness; and God saw that it was good. 19There was evening and there was morning, a fourth day.

It is clear that GOD thought of light and darkness and even created light BEFORE the sun, moon and stars. (as the discovery of cosmic background radiation proves.) Even formless matter, mass, water, BEFORE concentrating energy and mass into visible creation. GOD THOUGHT OF ALL THESE THINGS AND THEN MADE THEM MANIFEST. But I am not so particularly concerned personally about the order in which GOD THOUGHT of CREATION, but HIS

STATED REASONS FOR DOING SO. (I do not think of time or space only or exclusively in a linear fashion based on revelations I have seen; nor do I think all aspects of the Creation Account are past or have already been fulfilled; as I have already explained in previous citations.) So, what I want to focus on are the words: "let them be for signs and for seasons and for days and years". I already shared part of the spiritual connotations of various aspects pertaining to the fourth day in:

https://www.blastthetrumpet.org/PublicLetters/AAAUpdatedPublicAlertsMattersofLifeandDeath/Updates053016/Commentary%20on%20Genesis%20Part%204.pdf, but I want to point out WHAT CAN BE PLAINLY OBSERVED. The GOSPEL HAS BEEN TOLD IN THE STARS FOR ALL THE WORLD TO BEHOLD REGARDLESS OF THE LANGUAGE, TIME OR CULTURE OF THOSE PEOPLE WHO GAZE UPON THE CREATED STARS!

https://www.youtube.com/watch?v=BJ6szvfApNM THE STARS DIRECTLY CORRESPOND TO EVENTS ON EARTH AND PARTICULARLY IMPORTANT HISTORIC EVENTS! ESPECIALLY, the ANNUNCIATION, INCARNATION, CRUCIFIXION AND RESURRECTION OF CHRIST, THE LORD OUR GOD OUR SAVIOR!!!!!!!!
http://www.bethlehemstar.net/ BUT EVENTS DOWN THROUGH HISTORY ALSO HAVE BEEN DIRECTLY OBSERVED

AND LINKED TO THE CELESTIAL CLOCK, THE LIGHTS THAT OUR CREATOR MADE FOR DETERMINING THE TIMES, SIGNS AND SEASONS on earth!!!!!!!! Astronomical cycles have been DIVINELY ORDAINED and the wise know how they correlate to events on earth.

https://www.youtube.com/watch?v=EUQEMqF5dL8

https://www.youtube.com/watch?v=3n43opw6wiw

https://www.youtube.com/watch?v=f1p0so5jVw4 - side note on rebuttal of those who postulate a gap theory in the early verses of genesis

https://www.youtube.com/watch?v=HXUqfNQ3J0o

https://www.youtube.com/watch?v=YCsrveLNUDc

http://2.bp.blogspot.com/-BcEhkhQy9K4/Uz41Px68_4I/AAAAAAAABVo/yfUzj4J_61g/s1600/tetrads.png

http://www.bethlehemstar.net/what-does-this-mean/

http://2fletchdr222.blogspot.com/2014/04/the-history-of-8-blood-moon-tetrads.html

https://www.blastthetrumpet.org/PublicLetters/AAAUpdatedPublicAlertsMattersofLifeandDeath/Updates053016/End%20Times%20Verse%20By%20Verse.pdf

http://ipost.christianpost.com/news/the-four-blood-moons-in-history-and-next-12921/

https://www.youtube.com/watch?v=vi78PmRX46Y&list=PL29B760362587CED2&index=2

https://www.youtube.com/watch?v=PaKn8PI3B4Y&list=PL29B760362587CED2#t=254

http://www.skyscript.co.uk/kepler2.html - Kepler famous for his scientific findings in astronomy, noted that the cycles of the stars and conjunctions of planets were linked to historic events, even births of famous people; but rather than realize the DIVINELY CREATED CELESTIAL CLOCK, people whimsically disregard that aspect of Kepler's life under the notion of negative connotations regarding astrology. There is a difference between the fiction created in much of astrological texts, and those who through Wisdom Understand that Our CREATOR did indeed make the sun, moon and stars "to separate the day from the night, and let them be for signs and for seasons and for days and years..."

http://people.duke.edu/~ns2002/pdf/JohannesKepler.pdf - Just to emphasize that astronomy and astrology have always been linked historically. THE ERROR in astrology is FICTIONALLY creating and assigning aspects of personification to the planets and stars as if they were conscious gods bestowing blessings or cursings upon each soul and events on earth. WHAT IS TRUE, is that the

Constellations are a Celestial Clock that accurately determine days, years, seasons, and even correspond to important historic events on earth. (OBSERVABLE "signs" in the heavens) Divine Destiny, ordination of times and events, and even as shown, the very names of the stars tell of such events, particularly about our CREATOR and how to KNOW HIM; yet another way of how CREATION is OVERWHELMING EVIDENCE of our CREATOR.

http://www.hermetics.org/pdf/astrology/Johannes_Kepler_-_Concerning_The_More_Certain_Fundamentals_of_Astrology.pdf -

It is one thing to note that the sun, moon and even planets and constellations may affect the weather patterns on earth or coincide to weather patterns and quite another to claim people have certain personalities based on their birth time and should chart the course of their lives by fictional qualities attributed to celestial bodies. It is one thing to note that on a bright night or full moon people might have a more difficult time getting a good night sleep, and quite another to claim they turn into werewolves as a result. This is what I mean by discerning the difference between TRUE aspects and IMAGINARY ones; regarding connotations of astronomy and astrology. There is no doubt that Kepler studied both astronomy and astrology, but not in the popular horoscope fashion published in rag magazines of today. Rather he studied them as a scientist and a man of faith, looking objectively into any seeming connections of his astral observations to events in history and in his time. If the information presented in bethlehemstar.net DVD production is accurate and that "sign" is connected to CHRIST, THEN it would be reasonable to look for that "sign" again; according to:

http://biblehub.com/matthew/24-30.htm

The Return of the Son of Man ...29"But immediately after the tribulation of those days THE SUN WILL BE DARKENED, AND THE MOON WILL NOT GIVE ITS LIGHT, AND THE STARS WILL FALL from the sky, and the powers of the heavens will be shaken.30"And then the sign of the Son of Man will appear in the sky, and then all the tribes of the earth will mourn, and they will see the SON OF MAN COMING ON THE CLOUDS OF THE SKY with power and great glory. 31"And He will send forth His angels with A GREAT TRUMPET and THEY WILL GATHER TOGETHER His elect from the four winds, from one end of the sky to the other. (note "sky" can also be translated as "heaven" - http://biblehub.com/interlinear/matthew/24-30.htm) But for those of us who KNOW HIM, what will terrify those who REJECT HIM and DO NOT KNOW HIM, will only be assurance of HIS SOON GLORIOUS RETURN!!!!!!!! -

https://www.blastthetrumpet.org/PublicLetters/AAAUpdatedPublicAlertsMattersofLifeandDeath/Updates053016/The%20Climax%20of%20the%20Age%20Christ%20the%20End.pdf

So, the WORD OF GOD bears witness accurately of past, present and future, and CREATION itself confirms it. WHAT WE OBSERVE DIRECTLY CONFIRMS THE BIBLICAL ACCOUNT HISTORICALLY, NATURALLY, SUPERNATURALLY, PROPHETICALLY, UBIQUITOUSLY!!!!!!!!

https://www.youtube.com/watch?v=smepsIUEUrU

(someone should ask facebook WHY such information is unavailable (when digital data storage is becoming cheaper and cheaper these days) and to leave notes for posterity especially about matters of life, death and everlasting destinies - and then ask them to not have the appearance of trying to censor information that tells people how to KNOW and LEARN from our Eternal Creator personally now and forever; the most important information anyone who still doesn't know Him will ever read, see or hear)

http://www.blastthetrumpet.org/PublicLetters/AAAUpdatedPublicAlertsMattersofLifeandDeath/More

Updates031315/Evidence%20of%20GOD%20our%20Creator.pdf) - evidence of God our Creator.

https://www.blastthetrumpet.org/PublicLetters/AAAUpdatedPublicAlertsMattersofLifeandDeath/Updates053016/End%20Times%20Verse%20By%20Verse.pdf - End Times Verse by Verse.

https://www.blastthetrumpet.org/PublicLetters/AAAUpdatedPublicAlertsMattersofLifeandDeath/Prophecies%20Show%20Divine%20Inspiration%20of%20the%20Holy%20Bibl2.pdf - Prophecies show the Divine Inspiration of the Holy Bible.

https://www.blastthetrumpet.org/PublicLetters/AAAUpdatedPublicAlertsMattersofLifeandDeath/Updates053016/NO%20MORE%20BRAINWASHING%20p1.pdf - The Holy Bible is history. Its contents have been more thoroughly researched and verified over all other books in this world. Part 1

https://www.blastthetrumpet.org/PublicLetters/AAAUpdatedPublicAlertsMattersofLifeandDeath/Updates053016/NO%20MORE%20BRAINWASHING%20p2.pdf - There are more commentaries and apologetics on earth than any can read in their life time, that's how thoroughly researched and verified the contents of the Holy Bible are!   Part 2

https://www.blastthetrumpet.org/PublicLetters/AAAUpdatedPublicAlertsMattersofLifeandDeath/Updates053016/The%20Ten%20Commandments%20Part%201.pdf - Ten Commandment Part 1

https://www.blastthetrumpet.org/PublicLetters/AAAUpdatedPublicAlertsMattersofLifeandDeath/Updates053016/HOW%20TO%20KNOW%20TRUTH%20IN%20A%20WORLD%20OF%20LIES.pdf THE OBSERVABLE EVIDENCE SUPPORTING THE BIBLICAL ACCOUNT IS SO OVERWHELMING, ONE HAS TO CHOOSE TO BE WILLFULLY IGNORANT, NOT TO PERCEIVE REALITY!!!!!!!!

https://www.blastthetrumpet.org/PublicLetters/AAAUpdatedPublicAlertsMattersofLifeandDeath/Updates053016/Denial%20of%20REALITY.pdf - Remain true to God and His Words in the Holy Bible and you won't present yourself as those who are in denial of reality.

https://www.blastthetrumpet.org/PublicLetters/AAAUpdatedPublicAlertsMattersofLifeandDeath/Updates053016/Mentally%20Challenged.pdf - I may not be a smart man, but

I KNOW BEYOND ALL DOUBT THAT JESUS CHRIST IS LORD, THE ONE TRUE GOD. And so, I hope everyone would to their present and everlasting Joy!

https://www.facebook.com/photo.php?fbid=922427591169534&set=a.115635768515391.22520.100002069048072&type=3&theater They CHOOSE to be irrational and unreasonable, NOT scientific or reasonable by espousing and clinging to imaginations and fantasies in the name of science.

https://www.youtube.com/watch?v=5znVUFHqO4Q - historian Gary Habermas points out that people are NOT always rational and logical in their chosen beliefs; they don't ALWAYS base their chosen worldview on EVIDENCE; because if people did by now everyone on earth would KNOW our CREATOR personally.):

https://www.blastthetrumpet.org/PublicLetters/AAAUpdatedPublicAlertsMattersofLifeandDeath/Updates053016/Too%20Many%20are%20Turning%20Science%20Into%20a%20Religion.pdf In addition, those who falsely posture themselves upon imaginations in the name of science; the modern idolaters of religious acolytes of scientism HAVE TO IGNORE THE FINDING OF TRUE SCIENTISTS, the

GREATEST INTELLECTS and MOST POWERFUL PEOPLE IN ALL OF HISTORY!

https://www.blastthetrumpet.org/PublicLetters/AAAUpdatedPublicAlertsMattersofLifeandDeath/Updates053016/Fathers%20of%20the%20Sciences%20and%20Great%20Scientific%20Minds.pdf - Fathers of the Sciences and Great Scientific. The Age of Reason and Enlightenment is owed to God, Jesus Christ, and His Commandments in the Holy Bible. He taught to teach His Commandment to all children with each subsequent generation and so those content inspired the Gutenberg Printing Press with movable type which led to the printing of modern books and public education. Western Civilization owes its most noteworthy institutions to GOD, so make sure the contents of the Holy Bible remain as the standard of Truth in our public schools FOREVER!

They laugh and scoff at Biblical Creationists, when the VAST MAJORITY OF SCIENTIFIC NOBEL PRIZE WINNERS, are monotheistic Biblical Creationists! So, it is the evolutionists and atheists who desperately cling to vain imaginations and project themselves as if they are the rational, reasonable intellectuals; when the truth is THEY ARE THE RELIGIOUS ONES whose worldview is built upon BLIND BELIEFS; NOT the BIBLICAL CREATIONISTS!!!

https://www.blastthetrumpet.org/PublicLetters/AAAUpdatedPublicAlertsMattersofLifeandDeath/Updates053016/Evolution%20and%20Atheism%20Intertwined%20Cults%20of%20the%20Insane.pdf - While they laugh, mock and scoff at Biblical Creationists, the more they do so; the more it is THEY who are displaying themselves as truly

https://www.blastthetrumpet.org/PublicLetters/AAAUpdatedPublicAlertsMattersofLifeandDeath/Updates053016/Mentally%20Challenged.pdf!!!!!!!! Yes, I know I'm redundant, but the brainwashing has served to bring about a very detrimental collapse in our once gifted civilization, so I want our nation and all nations to learn to TRUST IN GOD AND HIS WORDS IN THE HOLY BIBLE for your own good!

BUT IF SOMEHOW there is any who has actually looked at all the information cited this far about HOW GOD OUR CREATOR confirms the Biblical Account NOT JUST PHYSICALLY, BY WHAT WE OBSERVE, BUT ALSO SPIRITUALLY, and NOT JUST GENERALLY, GLOBALLY, BUT specifically and personally, AND NOT JUST HISTORICALLY but PROPHETICALLY, and NOT JUST VISIBLY, BUT INVISIBLY, and NOT JUST EMPIRICALLY but LEGALLY, and NOT JUST MERCIFULLY but JUSTLY, and NOT JUST TERRESTRIALLY but CELESTIALLY, and not just in one field of scientific discipline BUT IN THEM ALL, but if SOMEHOW

YOU HAVE WATCHED THE HUNDREDS OF HOURS OF DOCUMENTARIES, READ THE LATEST SCIENTIFIC DISCOVERIES AND TAKEN A LOOK AT THE UNIVERSE AS CITED IN MY REFERENCES and STILL are in doubt as to WHO the ONE TRUE GOD IS! Then I encourage YET AGAIN to CALL UPON HIM!!!!!!!!

https://www.blastthetrumpet.org/PublicLetters/AAAUpdatedPublicAlertsMattersofLifeandDeath/Updates053016/You%20MUST%20Be%20Born%20Again.pdf - You Must be Born Again, of God's Spirit.

https://www.blastthetrumpet.org/PublicLetters/AAAUpdatedPublicAlertsMattersofLifeandDeath/Updates053016/GOD%20Part%201.pdf - God our Creator part 1

https://www.blastthetrumpet.org/PublicLetters/AAAUpdatedPublicAlertsMattersofLifeandDeath/Updates053016/GOD%20Part%202.pdf - God our Creator part 2

https://www.blastthetrumpet.org/PublicLetters/AAAUpdatedPublicAlertsMattersofLifeandDeath/Updates053016/Christ%20Crucified%20and%20Resurrected.pdf - Christ Crucified and Resurrected demonstrate not just how much God hates sin of all kinds, but also how much He loves us.

Since God determined that it was necessary for Him to show Himself and Prove Himself to us in this manner, every NEEDS to recognize Him for their own Good. Have a proper respect for God.

https://www.blastthetrumpet.org/PublicLetters/AAAUpdatedPublicAlertsMattersofLifeandDeath/Updates053016/BibleVquranTruthVlies.pdf - The Holy Bible is the literary foundation of TRUTH that God has given mankind by which all other opinions stand or fall upon, past, present and future. Anything spoken or written that so obviously goes against His Commandments in the Holy Bible, like the quran, IS NOT FROM GOD, but is from the lying devil!

And so we continue in Genesis with the Accurate Creation Account: The Fifth Day: Fish and Birds 20Then God said, "Let the waters teem with swarms of living creatures, and let birds fly above the earth in the open expanse of the heavens." 21God created the great sea monsters and every living creature that moves, with which the waters swarmed after their kind, and every winged bird after its kind; and God saw that it was good. 22God blessed them, saying, "Be fruitful and multiply, and fill the waters in the seas, and let birds multiply on the earth." 23There was evening and there was morning, a fifth day. http://biblehub.com/interlinear/genesis/1-21.htm -

As I read the text, here again it is apparent that dinosaurs and birds existed simultaneously. "Great sea monsters" has several connotations, one of which is sea dragons. The term "dragons" has historically and still in some places of the world currently refers to creatures that have been renamed under the heading of "dinosaurs". Our Creator PLAINLY tells us birds and marine life was created before land animals; BUT if the short time Creationists are correct, this order of creation could hardly be observed in the fossil record because land animals followed the very next day. So, if SHORT TIME (under 10000 years for the recorded history of mankind and everything in it is true), then the fossil record should be a jumbled mixture of ALL kinds of creatures devastated in the flood. (mammals, birds, dinosaurs, marine life, land animals, all over the world; and that is in fact what is found, contrary to the fictional "geologic column" espoused by so many through coerced indoctrination) Furthermore, it would make sense that the majority of the "dragons" that would have survived the flood would be marine creatures. Understand we have a historic period of "dragon hunters" all the way up into the Middle Ages.

http://www.nwcreation.net/dinosdragons.html

http://www.genesispark.com/exhibits/evidence/historical/ancient/dinosaur/

http://www.amnh.org/exhibitions/past-exhibitions/mythic-creatures/dragons-creatures-ofpower/european-dragons http://www.genesispark.com/exhibits/evidence/historical/dragons/ but crypto-zoologists have been documenting much more recent accounts.

https://www.youtube.com/watch?v=AUytp9LRVps

https://www.youtube.com/watch?v=T3S7xCnJ9bA&list=PLB5kkNKdCCsC0gFHZyMuT1jkTBJ43keQO

https://www.youtube.com/watch?v=T4csp6O5S-Y

https://www.youtube.com/watch?v=N6i45s_94A4

http://www.cryptozoology.com/forum/topic_view_thread.php?tid=5&pid=394941

https://www.youtube.com/watch?v=uRmTzWRy0eU

http://seamonster.org/

http://www.mysterycasebook.com/lakemonsterphotographs.html

https://www.google.com/search?q=japanese+fishermen+captured+a+dinosaur&sa=X&rlz=1C1GIGM_enUS535US535&espv=2&biw=1066&bih=702&tbm=isch&tbo=u&source=univ&ei=Td1mVMzJNIaRyQT3-4DQCg&ved=0CCQQsAQ

https://www.google.com/search?q=plesiosaur+washed+up+on+beach&rlz=1C1GIGM_enUS535US535&espv=2&biw=1066&bih=702&tbm=isch&tbo=u&source=univ&sa=X&ei=rN1mVPy2MoW0yATmkID4CQ&ved=0CB0QsAQ

https://www.youtube.com/watch?v=7IQv9mQLNGI

https://www.youtube.com/watch?v=y4JiITeVHy4

http://paleo.cc/paluxy/plesios.htm - of course skepticism will remain until incontrovertible evidence is found, but for many that would mean seeing one up close first hand. http://www.strangemag.com/seaserpgallery.html - One thing I have noted is that for many people the mystery is more exciting than the answers.

https://www.facebook.com/photo.php?fbid=439607769532636&set=gm.979771358730522&type=1&t heater Notwithstanding healthy skepticism, I personally think the evidence is overwhelming that THERE IS MUCH unexplained and undiscovered species still all over our world AND that dinosaurs or popular connotations of "prehistoric" creatures that are STILL WITH US are only considered by other names BECAUSE they are still ALIVE and not extinct. Just consider the attributes of these species:

https://www.google.com/search?q=giant+maneating+crocodile+captured+in+uganda&rlz=1C1GIGM_enUS535US535&espv=2&biw=1066&bih=702&tbm=isch&tbo=u&source=univ&sa=X&ei=reFmVN3ABc71yATJyoD4Cg&ved=0CB0QsAQ&dpr=1.5

https://www.youtube.com/watch?v=7IYMK22QCMA

https://www.youtube.com/watch?v=q5cnvkHYqLg

https://www.google.com/search?q=indonesia+massive+snake&rlz=1C1GIGM_enUS535US535&espv=2&biw=1066&bih=702&source=lnms&tbm=isch&sa=X&ei=4-JmVP_GBtaoyASW2oHgBw&ved=0CAYQ_AUoAQ

https://www.youtube.com/watch?v=xeOSXtBCY30

https://www.youtube.com/watch?v=NzzxOlFJtzg

https://www.youtube.com/watch?v=mneDhOtVEQw

https://www.youtube.com/watch?v=Pa0SY6RYJlo

https://www.google.com/search?q=rhinoceros&espv=2&biw=1066&bih=702&source=lnms&tbm=isch&sa=X&ei=0KVxVIXmNZDHsQT1kIHoAQ&ved=0CAYQ_AUoAQ&dpr=1.5

https://www.youtube.com/watch?v=aoHxdwsLxos

https://www.youtube.com/watch?v=ieR7hUHs4GU

https://www.google.com/search?q=pangolin&espv=2&biw=1066&bih=702&source=lnms&tbm=isch&sa=X&ei=CaRxVIauHM3fsATy4IHoBA&ved=0CAYQ_AUoAQ

https://www.google.com/search?q=armadillo&espv=2&biw=1066&bih=702&source=lnms&tbm=isch&sa=X&ei=76RxVKKtGM7ksASa24HgBw&ved=0CAYQ_AUoAQ&dpr=1.5

http://voices.nationalgeographic.com/2013/11/08/bizarre-mystery-fish-identified-freshwater-speciesof-the-week/

https://www.youtube.com/watch?v=i9pJ-JHYayc - I include video evidence like this to show various species STILL alive and existing with diverse other species SIMULTANEOUSLY are exactly the kind of species evolutionists claim are "transitory forms"; rather than what we can plainly see. (Diverse species simultaneously existing all together currently and in the past).

https://www.youtube.com/watch?v=ntZxw6N-5oM - We see SIZE variations of ALL KINDS OF SPECIES LARGER in the past than the present. So, imagine what sharks, crocodiles, alligators and various species like these looked like not so long ago.

https://www.youtube.com/results?search_query=komodo+dragon+attacks

https://www.youtube.com/results?search_query=lizards+that+run+on+two+legs

https://www.youtube.com/watch?v=RIb72ok14wE

https://www.google.com/search?q=giant+fossil+cockroach&rlz=1C1GIGM_enUS535US535&espv=2&biw=1066&bih=702&source=lnms&tbm=isch&sa=X&ei=QPJmVODPKo7-_yQSd1YDwCA&ved=0CAYQ_AUoAQ&dpr=1.5

https://www.google.com/search?q=giant+fossil+cockroach&rlz=1C1GIGM_enUS535US535&espv=2&biw=1066&bih=702&source=lnms&tbm=isch&sa=X&ei=QPJ

mVODPKo7-yQSd1YDwCA&ved=0CAYQ_AUoAQ&dpr=1.5#tbm=isch&q=giant+fossil+insects

My point here is that in the not too distant past, creatures that are common and STILL ALIVE TODAY were simply larger and as such would have easily qualified as "dinosaurs" or other "prehistoric" creatures. MANY OBSERVATIONS of species to date follow that pattern and under certain theories as to why, some creationists have caused common species of today to grow abnormally large.

https://www.facebook.com/notes/god-in-science/the-evolutionary-tree-isimaginary/892988044118276

http://www.dailymail.co.uk/sciencetech/article-3117137/Blood-skin-cells-75-million-year-old-dinosaurbones-Tissue-extracted-fossils-left-storage-century.html dinosaur soft tissue samples are PROLIFIC compared to the missing links of darwinian evolution which are non-existent. http://kgov.com/dinosaursoft-tissue truth is there are NO KNOWN conditions that account for such preservation over the FALSELY ASSUMED millions of years to hundreds of millions of years.

https://www.youtube.com/watch?v=AjheQjMwHJA

http://www.rexresearch.com/baugh/baugh.htm

http://www.creationevidence.org/evidence/crystillane_canopy_theory.php

http://discourseanddisclosure.com/frontPage/tomato.htm

http://www.genesispark.com/exhibits/early-earth/experiments/

http://danieltrainingnetwork.org/glory-original-earth-biblical-theology-of-mission/

https://www.youtube.com/watch?v=Aq4DK4W9xe4#t=97

Catastrophic flood (NOT "evolution") is the rational explanation for what we observe in fossils, artifacts and history.

https://www.facebook.com/photo.php?fbid=922427591169534&set=pb.100002069048072.-2207520000.1457756152.&type=3&theater

https://www.youtube.com/watch?v=yH35OQbV7GY

https://www.youtube.com/watch?v=hmMIpgIyOnY

https://www.youtube.com/watch?v=eJsAn2jSxxo

https://www.facebook.com/1mill.creationist/photos/a.361018770595969.88428.286013238096523/875899082441266/?type=1&theater - philology traces back to the ark location where it came to rest following the receding waters.

Noah planted an orchard and its right where it should be: https://www.youtube.com/watch?v=IoTkguzRaCU

http://www.soundchristian.com/man/

and

http://www.scribd.com/doc/22885188/Hamitic-Semitic-Japhetic

and various artifacts all over the world confirm.

http://www.econ.ohio-state.edu/jhm/arch/batcrk.html

https://www.google.com/search?q=paleo+hebrew+ancient+artifacts&espv=2&biw=1066&bih=702&tbm=isch&tbo=u&source=univ&sa=X&ei=4C9sVNu8K4aAygTd7IC4Cw&ved=0CB0QsAQ Some people still roll their eyes about the flood; when the evidence for it is so overwhelming it should be common knowledge by now.

Common ridiculous arguments about the size of Noah's Ark and number of species usually are given while ignoring all the evidence that sea worthy craft of that size and date have been found in the region as I cover in

https://www.blastthetrumpet.org/PublicLetters/AAAUpdatedPublicAlertsMattersofLifeandDeath/Updates053016/Evidence%20of%20the%20Flood.pdf and

https://www.blastthetrumpet.org/PublicLetters/AAAUpdatedPublicAlertsMattersofLifeandDeath/Updates053016/NO%20MORE%20BRAINWASHING%20p1.pdf - part 1

https://www.blastthetrumpet.org/PublicLetters/AAAUpdatedPublicAlertsMattersofLifeandDeath/Updates053016/Bill%20Nyes%20Science%20Lies.pdf but something I don't often hear skeptics acknowledge are certain facts like not that many kinds of animals would have needed to be on board even in a global flood.

http://www.palaeontologyonline.com/articles/2014/fossil-focus-arthropod-plant-interactions/

http://www.nhnct.org/fauna/fauna.html

http://gianthamster.com/wp-content/uploads/2010/05/TypesOfAnimals.jpg

http://burgess-shale.rom.on.ca/en/science/burgess-shale/03-fossils.php

As anyone can see a VERY SMALL percentage of all known species would have needed to be on the Ark to survive the global flood. Of which the vast majority of those could be preserved as seeds, or are so small as to be negligible. Furthermore, consider the specific instructions: Entering the Ark 1Then the LORD said to Noah, "Enter the ark, you and all your household, for you alone I have seen to be righteous before Me in this time. 2"You shall take with you of every clean animal by sevens, a male and his female; and of the animals that are not clean two, a male and his female; 3also of the birds of the sky, by sevens, male and female, to keep offspring alive on the face of all the earth....

So, a) we can see the vast majority of species on earth had no need to even be on the Ark and b) there is more than just single breeding pairs of many species of those that were on the Ark to account for present day populations. Has anyone who argues against the Ark seriously studied procreation rates of the species that would have been on board?

http://creation.com/where-are-all-the-people

http://creation.com/Flood-biogeography

Population rates of species accurately account for a global bottleneck at the time of the historic global flood.

https://www.youtube.com/watch?v=o0ghHia-M54 -

This mathematician clearly counters deep time claims by evolutionists. (As populations would have long ago depleted the natural resources on earth and caused widespread extinction events). The "doubling time" of various species including our own corresponds directly with the date of the global flood. But other scientists do argue against these notions by pointing out at least for humans that birth and death rates (as pertaining to population growth) fluctuate and are on the overall decline:

https://www.youtube.com/watch?v=eA5BM7CE5-8

Regardless of the views, the math and recorded history ACCURATELY places the "bottleneck" of humanity; from "mitochondrial" Adam and Eve to the generation of Noah, at the date of the flood; MOST DEFINITELY NOT "hundreds of thousands" or even "millions of years ago"; as yet ANOTHER FACT proving modern so called radiometric dating methods are RUBBISH and the popular evolutionary paradigm totally bunk!

http://www.cs.unc.edu/~plaisted/ce/dating2.html
http://christiannews.net/2015/07/27/science-supports-genesis-geneticist-says-evidence-confirmshistorical-adam-and-eve/ -

Even the most conservative population growth rates would have filled this world billions of times over, if mankind and other species have been around as long as evolutionists claim. RATHER population doubling rates correspond to the bottleneck mitochondrial Adam and Eve exactly with the Biblical timeframe of the Creation Account followed by the Global Flood.

http://www.blastthetrumpet.org/PublicLetters/AAAUpdatedPublicAlertsMattersofLifeandDeath/MoreUpdates041215/The%20religion%20of%20evolution%20versus%20the%20Science%20of%20Creation.pdf - the cultic views of evil-u-shun verses the SCIENCE OF BIBLICAL CREATION.

http://www.blastthetrumpet.org/PublicLetters/AAAUpdatedPublicAlertsMattersofLifeandDeath/MoreUpdates031315/Evidence%20of%20GOD%20our%20Creator.pdf - Evidence of God our Creator.

http://answersinscience.org/RadiometricDating-Woolf.htm - those claiming "deep time" geologic formation DESPITE the ubiquitous evidence that it is FALSE, point to radiometric dating as their "STRONGEST evidence" in support of their claims. Yet we can see their "strongest evidence" is NO EVIDENCE at all!!!!!!!!

http://creationtoday.org/radiometric-dating-is-it-accurate/ - it is the ASSUMPTIONS that cause the dates to be WRONG.

https://www.youtube.com/results?search_query=hovind+radiometric+dating+fallacies - you can look and decide for yourselves because people on both sides of the issue are outspoken.

http://www.christiananswers.net/q-aig/aig-c007.html - I, personally, have yet to be convinced of the age of the universe or the earth having spent THOUSANDS of hours of research to both sides of the argument. I AM FIRMLY CONVINCED AND KNOW BEYOND ALL DOUBT OUR ETERNAL CREATOR CAN DO ANYTHING HE CHOOSES TO, the exception being He has stated He does not lie.

Instantaneous creation of the universe from His own existence is absolutely within His Power.

http://biblehub.com/matthew/19-26.htm but the modern lunacy coming out of theoretical physics these days of something from nothing is contradictory to sound knowledge and even the foundational precepts of epistemology. (something from nothing immediately violates the Laws of Non- contradiction and Causality - proving anyone making such a statement is irrational, illogical, and quite possibly insane). I have yet to see any PROOF from all those debating the issues, even from the most famous of scientists, for a universe of "billions of years"; whereas the "young universe" primarily relies on the authority of their understanding of the scriptures and some scientific observations; hotly debated by those claiming deep time, of course.

https://answersingenesis.org/evidence-for-creation/the-10-best-evidences-from-science-that-confirma-young-earth/

http://www.icr.org/article/evidence-for-young-world/

http://rationalwiki.org/wiki/101_evidences_for_a_young_age_of_the_Earth_and_the_universe

https://www.facebook.com/icr.org/photos/a.10150268208564451.345328.43552419450/10152911405884451/?type=1&theater

https://www.google.com/webhp?sourceid=chrome-instant&ion=1&espv=2&ie=UTF8#q=polonium%20halos%20unrefuted%20evidence -

I AM MOST DEFINITELY; BEYOND ALL DOUBT, A BIBLICAL CREATIONIST, but I have yet to be persuaded, as to the WHEN it all began. And frankly, think it more of an intellectual exercise, and spiritual quest, than the current events that need IMMEDIATE SERIOUS ATTENTION WORLDWIDE. Since God tells us he is in all, through it all, and above all, something Eternal is ever present with us, so I think we cannot tell the date of creation just by observing it, rather we must decide on that through asking our Creator and the study of His Word. We are facing FAR MORE URGENT MATTERS of global extinction events; on the verge of massive droughts, famines and wars, and yet billions of souls STILL do not know the ONE TRUE GOD; while people are debating the past in which no one on earth has firsthand knowledge of; unless our Creator showed them the TRUTH in that regard by Divine Revelation. GOD is ETERNAL, we might be able to trace recorded history back all the way to Adam and Eve, but the ages or epochs prior to His First Recorded Thought I cannot find in the HOLY BIBLE, just when was the FIRST THOUGHT of GOD? HE IS ETERNAL without Beginning and

End!!!!!!!! As such, I do not see in the HOLY BIBLE anywhere where He tells us how long He has existed (other than ETERNALLY without Beginning or End; but I mean from our connotations of time; years, how many have elapsed since His first thought of Creation) or for that matter how long the visible manifestation of all creation has existed. I DO see details of recorded history from Adam and Eve; especially, after the fall. And yet even in that history, I do not see prior to the MANIFESTATION of Adam and Eve, specific time frames for the Thoughts of GOD. I have discussed my views of the "days" of creation herein and in my citations already. As far as I'm concerned mankind should hold of paramount importance the creation of Adam and Eve and our HISTORY as opposed to any musings about PREHISTORY that so many seem to be engaged in.

https://www.blastthetrumpet.org/PublicLetters/AAAUpdatedPublicAlertsMattersofLifeandDeath/Updates053016/Commentary%20on%20Genesis%20Part%201.pdf - Commentary on Genesis part 1

I seek TRUTH and in that seeking I find PRIORITIES, and the PRIORITIES of our CREATOR are plainly stated!

https://www.blastthetrumpet.org/PublicLetters/AAAUpdatedPublicAlertsMattersofLifeandDeath/Updates053016/HOW%20TO%20KNOW%20TRUTH%20IN%20A%20WORLD%

20OF%20LIES.pdf - You must know the One who is Truth, Jesus Christ, to know Truth in this present world of so many lies.

https://www.blastthetrumpet.org/PublicLetters/AAAUpdatedPublicAlertsMattersofLifeandDeath/Updates053016/Dont%20Fail%20to%20Savor%20the%20Best%20Part.pdf - Once you KNOW the Creator, make certain everyone else does also!

https://www.blastthetrumpet.org/PublicLetters/AAAUpdatedPublicAlertsMattersofLifeandDeath/Updates053016/Creation%20V%20Evolution%20Use%20Your%20Intelligence%20to%20Recognize%20INTELLIGENCE.pdf - Intelligence is overwhelming evidence of our Intelligent Eternal Creator.

https://www.blastthetrumpet.org/PublicLetters/AAAUpdatedPublicAlertsMattersofLifeandDeath/Updates053016/No%20Excuses.pdf - No one has any excuse. Everyone can KNOW our Eternal Creator beyond all doubt.

https://www.blastthetrumpet.org/PublicLetters/AAAUpdatedPublicAlertsMattersofLifeandDeath/Updates053016/Set%20Apart.pdf - Those who wisely choose to follow and obey our Lord Jesus Christ, are to remain set apart for him so that they are a shining light to those still in darkness.

https://www.blastthetrumpet.org/PublicLetters/AAAUpdatedPublicAlertsMattersofLifeandDeath/Updates053016/Soldiers%20of%20Christ%20First%20General%20Order.pdf - I have reasoned that we can follow God's Commandments in Heaven, EXCEPT FOR THE GREAT COMMISSION! So, your number one purpose in Loving God and Loving Each Other is to make certain you tell everyone HOW TO KNOW HIM! (There aren't any unbelievers in Heaven, so make sure you do your best to make certain everyone knows that our LORD JESUS CHRIST is the Way there!)

Our Eternal Creator, LORD and SAVIOR, knows we love to learn and explore, HE MADE US, but we have ALL ETERNITY to search out the mysteries of His Creation. HE ASKS US TO FOCUS RIGHT NOW ON MAKING SURE EVERYONE ON EARTH KNOWS HIM AND IS LEARNING FROM HIM DIRECTLY!!!!!!!!

https://www.blastthetrumpet.org/PublicLetters/AAAUpdatedPublicAlertsMattersofLifeandDeath/Updates053016/Life%20or%20Death.pdf - most everything I have written on are about matters of life, death and everlasting destinies. I hope you take time to read them.

https://www.blastthetrumpet.org/PublicLetters/AAAUpdatedPublicAlertsMattersofLifeandDeath/Updates053016/Choose%20LIFE%20Avoid%20Death%20and%20Destruction.pdf - choose Life; choose JESUS CHRIST!

https://www.blastthetrumpet.org/PublicLetters/AAAUpdatedPublicAlertsMattersofLifeandDeath/Updates053016/Yes%20But%20What%20Can%20I%20Do.pdf - We all have Divine Purposes, just ask the Divine for what yours is and concentrate on the His Greatest Commandments and the Great Commission with your lives!

The PRIMARY REASON I am addressing the brainwashing going on in public education is because it is leading mass millions AWAY from TRUTH, away from our CREATOR who is TRUTH and has become exceedingly evil and corrupt to the point where it MUST be addressed and the fallacies OVERWHELMINGLY EXPOSED such that we restore the proper TEACHER, GOD, to public education worldwide. These deceptions ARE INHERENTLY DESTRUCTIVE; not just to those espousing them, but to the WHOLE WORLD. (Just look at what has happened to our nation and the world since satanists (anyone showing adversity to God and His Words in the Holy Bible) took control of finance, media and education)

http://www.blastthetrumpet.org/PublicLetters/THE%20PLANNED%20DEMORALIZATION%20OF%20AME RICA.pdf - Counteract this agenda of Demoralization of our Nation with the Gospel of Jesus Christ and to put the Holy Bible back in our schools and let every child be taught from it throughout all their years of learning.

https://www.blastthetrumpet.org/PublicLetters/AAAUpdatedPublicAlertsMattersofLifeandDeath/Updates053016/Brainwashing.pdf - never let lies in the name of science so pervert our society. The HOLY BIBLE IS THE TRUTH; it needs to be taught instead of these fallible notions from fallible people!

https://www.blastthetrumpet.org/PublicLetters/AAAUpdatedPublicAlertsMattersofLifeandDeath/Updates053016/Conspiracy%20Theories%20or%20Plain%20Truth.pdf - conspiracy theories or plain truth. There are indeed conspiracies taking place today and are perverting and corrupting our nation.

https://www.blastthetrumpet.org/PublicLetters/AAAUpdatedPublicAlertsMattersofLifeandDeath/Updates053016/Treatise%20on%20Greed%20and%20Corruption.pdf - Greed is definitely one of the ways corruption is entering our society.

https://www.blastthetrumpet.org/PublicLetters/AAAUpdatedPublicAlertsMattersofLifeandDeath/Updates053016/PROPAGANDA.pdf - Propaganda is another way. People must know the Truth (Jesus Christ) in order to not be swayed by all these devious methods.

https://www.blastthetrumpet.org/PublicLetters/AAAUpdatedPublicAlertsMattersofLifeandDeath/Updates053016/satanil%20v%20GOD%20part%202.pdf - satanil, the deceiver, is at work deceiving the minds of people who do not KNOW JESUS CHRIST and is why they are resistant to the Gospel Message COMMANDING EVERYONE TO REPENT OF ALL SINS and to GET BAPTIZED in the Name of our LORD JESUS CHRIST, so they can be filled with God's Holy Spirit and KNOW HIM, here and now and forever.

https://www.blastthetrumpet.org/PublicLetters/AAAUpdatedPublicAlertsMattersofLifeandDeath/Updates053016/Depopulation.pdf - Depopulation, by intent or accident. Our culture has depopulation agendas that make people think they are doing the right thing but can actually be harming people. You must be very certain of your health and do your best to maintain it lifelong.

https://www.blastthetrumpet.org/PublicLetters/AAAUpdatedPublicAlertsMattersofLifeandDeath/Updates053016/Strategies%20of%20War.pdf - When we think of war, we often only think militaristically but warfare is accomplished in a variety of ways, the general population needs to recognize the hostility arrayed against them by some of the greediest people on earth. There mannerism of maximum profits is having an adverse effect on the general population.

Their lies and deceptions intentionally perverting innocent children are causing widespread corruption, vices, and worse overt devil worship to invade our land which is leading to the destruction not just of our liberties and economies, but lives!

https://www.blastthetrumpet.org/PublicLetters/AAAUpdatedPublicAlertsMattersofLifeandDeath/Updates053016/End%20Times%20Verse%20By%20Verse.pdf - End times, verse by verse.

https://www.blastthetrumpet.org/PublicLetters/AAAUpdatedPublicAlertsMattersofLifeandDeath/Updates053016/The%20Sufferings%20of%20Christ.pdf - There are still persecution by the ungodly against innocent Christians all over this world. The sufferings of Christ are still present today! So, when He says He is coming back with a Vengeance to all who fail to repent, understand He will be out of patience with all ungodly persons!

2 Thessalonians 1:8-9

King James Version

8 In flaming fire taking vengeance on them that know not God, and that obey not the gospel of our Lord Jesus Christ:

9 Who shall be punished with everlasting destruction from the presence of the Lord, and from the glory of his power;

https://www.blastthetrumpet.org/PublicLetters/AAAUpdatedPublicAlertsMattersofLifeandDeath/Updates053016/Observable%20Divine%20Judgments.pdf - There are already observable Divine Judgments the more we depart from Him, read Deut 28 and see the consequences are already upon our nation and the world for departing from our Eternal Creator. We must return to Him zealously and never depart again. We can see His Consequences in all the earth!

So, I am addressing these perversions of science as the lies that they are BECAUSE they are PROVABLY destroying our nation, our heritage, our freedoms, and if not stopped leads to oppression, open wickedness, death and destruction. BUT it is NOT JUST present death and destruction that concerns me; it is EVERLASTING DESTINIES!!! I have been "beyond and back" myself. I KNOW, BEYOND ALL DOUBT, the Biblical Account of Existence after "death" is TRUE!!!!!!!! As such, I MUST do everything in my power to make certain every man, woman and child on earth KNOWS our ETERNAL CREATOR, LORD and SAVIOR, JESUS CHRIST, to their present and everlasting Joy!!!!!!!! So, to continue with how what we observe supports what is plainly recorded in the Holy Bible and does NOT support evolutionary fiction in any way:

http://creation.com/in-pelegs-days-the-earth-wasdivided -

In all my study of God's Word I have found He says but one word and yet it refers to MULTIPLE ASPECTS of His Creation, physical, spiritual, general, specific; and so here in Genesis, I think based on the evidence of huge canyons and crustal displacements worldwide, that not only was the Global Flood of Noah's Day EPIC and massively cataclysmic, but also the separation of the continents in RECENT history as well! (I said it before and I'll say it again, I distinctly remember being shown a film strip in grade school that showed a tree in South America that had been split and the other half in Africa and it was shown us as absolute proof that the two continents were at one time joined together.) At the time I raised my hand because the teacher was saying that the separation of the two continents occurred "millions of years" ago, and so I asked if the presentation was stating that split tree was "millions of years old"? She said that was a good question and never answered it. My point is that they HAVE found marsupials fossilized in Egypt contrary to those who roll their eyes about the global flood being actual history.

http://karionghieroglyphs.blogspot.com/ and as previously discussed marsupials have been found elsewhere besides Australia (on virtually every continent) and present dating methods are wildly inaccurate. In addition, mankind obviously had knowledge of ocean-going vessels and so it is not necessarily a fact that all species migrating after the flood utilized land bridges, people and animals, seeds etc. could have traveled by sea; not to mention huge migrations of birds and insects; together with whatever they carried with them on or in their organisms. Also remember that it is far more likely that the vast majority of all fossils found would be what existed PRIOR to the flood; NOT after. (natural decomposition predominantly occurs apart from catastrophic events) AND with the BABEL account and the destruction of the tower, we have another explanation for what caused the global dispersion of both humans and animals.

http://www.gotquestions.org/Tower-of-Babel.html and

http://www.creationdefense.org/34.htm

I believe that the evidence points not only to migration as a possible explanation but also a more recent and rapid continental separation not so much by plate tectonics as rather a change in land elevations due to the flood creating new emergences of "continents"; while what was previously above water, much still remains below. (The Flood Epic, corresponded to major geologic changes on the surface of the earth, like all the canyons; even deep-sea canyons, opening up all over the world as the possible explanation for the receding waters). Regardless, of not finding fossils of every species YET; currently alive today, we do have evidence of recent historic discoveries that show ancient connections between even far away Australia to the Middle East.

http://www.bibliotecapleyades.net/esp_australia02.htm

http://wakeup-world.com/2014/10/14/hieroglyphics-experts-declare-ancient-egyptian-carvings-inaustralia-authentic/

http://www.mysteriousaustralia.com/egyptians_australia_mainpage.html And other artifacts link the Middle East even to the Americas:

http://www.in-the-desert.com/giants.html

https://www.google.com/search?q=golden+plates+found+in+utah&sa=X&rlz=1C1GIGM_enUS535US535&espv=2&biw=1066&bih=702&tbm=isch&tbo=u&source=univ&ei=1y1nVNGLPISoyQTR6IBo&ved=0CCc QsAQ

http://projectavalon.net/lang/en/klaus_dona_2_interview_transcript_en.html - understand that dating methods have been proven entirely inaccurate and/or incorrect because of false assumptions; but what the presentation clearly does prove is pre-flood and post-flood global civilization.

https://www.youtube.com/watch?v=lFHwKmHfdo8

https://answersingenesis.org/fossils/how-are-fossils-formed/do-fossils-show-signs-of-rapidburial/?utm_source=facebook-aig&utm_medium=social&utm_content=dofossilsshowsigns16232&utm_campaign=20141117

Evidences of the global flood, arks (remnants of ships fitting dimensions and regional location of the Biblical Account), cities still beneath oceans and lakes, fossils gasping for air, clustered fossil graveyards, polystrate fossils, origins of human civilization and language and species migration from earliest written records, geologic formations and more. Recap on the Global Flood Noah's Ark and what followed:

1) Remnants of ships with evidence of animals have been found in the region indicated in the HOLY BIBLE -

https://www.blastthetrumpet.org/PublicLetters/AAAUpdatedPublicAlertsMattersofLifeandDeath/Updates053016/NO%20MORE%20BRAINWASHING%20p1.pdf)

The geologic global flood layer(s) is a FACT:

https://www.youtube.com/watch?v=ZnfXvPZhjfs

https://www.youtube.com/watch?v=7Pth1EC5N0E

https://www.youtube.com/watch?v=2Mt1c_OB_cM

https://www.youtube.com/watch?v=jJ1QrbvBIN4

we saw plainly these layers CANNOT BE "millions of years old":

https://www.youtube.com/watch?v=TWINTLPozMo

https://www.google.com/search?q=polystrate+fossils&rlz=1C1GIGM_enUS535US535&espv=2&biw=106

6&bih=702&tbm=isch&tbo=u&source=univ&sa=X&ei=aZxsVJqIKlayyQSJkIDADA&ved=0CCwQsAQ

https://www.youtube.com/watch?v=hqUtY23L4iQ

we saw experiments showing rapid layers can form in sediments WITHOUT deep time:

https://www.youtube.com/watch?v=DY7Wsf1T7bc

https://www.youtube.com/watch?v=Fu38VVfUYLk

https://www.youtube.com/watch?v=3-Ju2QJ2Q_o

https://www.youtube.com/watch?v=J48fLpiFHug

3) OLDEST WRITTEN RECORDS and oratory history on each continent TELLS of the global flood.

https://www.youtube.com/watch?v=lktmmd7YnD8

http://www.talkorigins.org/faqs/flood-myths.html

4) STILL BURIED CITIES, CIVILIZATIONS, beneath oceans and lakes:

https://www.google.com/webhp?sourceid=chromeinstant&rlz=1C1GIGM_enUS535US535&ion=1&espv=2&ie=UTF-8#q=civilizations%20underwater

https://www.youtube.com/results?search_query=sunken+civilizations+all+over+the+world

https://www.youtube.com/results?search_query=pyramids+underwater+all+over+the+world

https://www.youtube.com/results?search_query=cities+under+lakes

https://www.google.com/webhp?sourceid=chromeinstant&rlz=1C1GIGM_enUS535US535&ion=1&espv=2&ie=UTF-8#q=preflood%20civilizations%20now%20still%20underwater

5) Origins of languages, human migration, procreation rates all as documented in the HOLY BIBLE and as OBSERVED points clearly to a historic global flood in which many "dinosaurs" and even real "giants" and ancient civilizations were destroyed. (Mesopotamian region known as “the cradle of civilization”.)

https://www.google.com/webhp?sourceid=chromeinstant&rlz=1C1GIGM_enUS535US535&ion=1&espv=2&ie=UTF8#q=giant%20human%20skeleton%20found%20underwater

https://www.google.com/search?q=giant+human+skeleton+and+fossilized+dinosaurs+found+underwater&rlz=1C1GIGM_enUS535US535&oq=giant+human+skeleton+and+fossilized+dinosaurs+found+underw

ater&aqs=chrome..69i57&sourceid=chrome&es_sm=93&ie=UTF-8

Scientists have also discovered canyons even deep-sea canyons and huge underground oceans that can EASILY account for "where did all the water go?"

https://www.blastthetrumpet.org/PublicLetters/AAAUpdatedPublicAlertsMattersofLifeandDeath/Updates053016/Bill%20Nyes%20Science%20Lies.pdf - people think they mean well but when they learn lies and pass them on, they can harm others in the process. We need to teach children the Truth, FROM THE HOLY BIBLE.

https://answersingenesis.org/the-flood/how-could-fish-survive-the-genesisflood/?utm_source=facebook-aig&utm_medium=social&utm_content=howcouldfishsurvive16280&utm_campaign=20141120 - during floods, debris, much debris floats; even a cataclysmic flood as described in the Biblical account would have had such matter and debris, seeds, and small life forms, and of course marine species that could have easily survived and contributed to current populations today.

https://www.youtube.com/watch?v=lktmmd7YnD8&list=PLb-aLQEd-g8bXMvN9NY5DxTS7aQPJsjhg

I HAVE JUST SHOWN PLAINLY THAT THE UBIQUITOUS EVIDENCE, KNOWLEDGE OF MANKIND TO DATE, PLAINLY SUPPORTS THE GENESIS CREATION ACCOUNT AND EVEN THE MOST HOTLY DEBATED ASPECTS OF IT! At least, there is FAR MORE OBSERVABLE EVIDENCE in support of the Biblical Account of Creation, than for the nonsense being taught in public education these days! (namely, EVERYTHING WE OBSERVE SUPPORTS THE BIBLICAL ACCOUNT OF CREATION; conversely, NOTHING we observe supports the popular evolutionary paradigm being forced on mass millions in the name of so-called science)!

https://www.youtube.com/watch?v=HaQqa90e9KQ - perhaps the most evolutionary of all things on earth is how much the darwinian theory keeps changing in a vain effort to try and be inclusive of all the evidence that has disproven darwin's theory many times over. I address those who snicker about Eve being made from the rib of Adam (even though not so intelligent (compared to our ETERNAL CREATOR) and extremely fallible people have demonstrated once they have the hardware, living cells, and the software, DNA, they can bioengineer organs -

https://www.facebook.com/video.php?v=1015202922937 2917; yet somehow, that is supposed to be difficult for the Eternal Creator who brought the entire universe into existence), and other objections in my notes: help people not to present themselves as

https://www.blastthetrumpet.org/PublicLetters/AAAUpdatedPublicAlertsMattersofLifeandDeath/Updates053016/Mentally%20Challenged.pdf by teaching the Truth!

https://www.facebook.com/photo/?fbid=1380774058668216&set=a.115635768515391 - God is REAL! and https://www.facebook.com/photo/?fbid=977229375689355&set=a.115635768515391 - When you examine the evidence for God our Creator, it's overwhelming! Everyone should know Him already! https://www.facebook.com/photo/?fbid=969688109776815&set=a.115635768515391 - Jesus Christ is by far and away the most documented person on earth! Those mythologizing him are not in their right minds, they're so deceived. No one has impacted Creation like our Eternal Creator!

https://www.blastthetrumpet.org/PublicLetters/AAAUpdatedPublicAlertsMattersofLifeandDeath/Updates053016/The%20Sufferings%20of%20Christ.pdf and even though the

HOLY BIBLE makes it clear that the tree of the knowledge of evil is a metaphor for the deceptive spirit known as satanil; AND even though serpents are also clearly metaphors of wicked people.

http://biblehub.com/matthew/23-33.htm, I will address the "talking serpent" issue a little more in depth. I want to address this more particularly because these end times are bringing into manifestation open devil worshipers and so I call all those loyal to our Creator to arms; for the time is upon us when there will be open warfare all over the globe between all forms of satanism and those who are Faithful to our Eternal Creator, the ONE TRUE GOD. In the Divine Revelations I saw, it was as if I were looking through spiritual x-ray glasses, that enabled me to look at souls, not just outward appearances. And I SAW that even though OUTWARDLY people look alike or similar, INWARDLY angels and devils are VERY different! The serpent seed is spread abroad all the earth and are declared adversaries of GOD ALMIGHTY and His Followers. Inwardly, spiritually, the souls of these devils ACTUALLY LOOK reptilian or serpentine! (Hence all the "alien" drama surrounding these terms). They DANCE serpentine, they have orgies, like serpents, they worship serpents, indiscriminate sex with whoever, whatever, are factually devil worshippers that trace back to the same activities that brought about the flood of Noah's day and the destruction of Sodom and Gomorrah. I think that there is evidence that in the past these reptilians or walking, talking serpent people actually PHYSICALLY resembled such and that over time, through genetic engineering and modification have infiltrated our world to OUTWARDLY look like other human beings, while INWARDLY are the spawn of devils! Understand that is what the Battle of Armageddon is REALLY about. It's not just about the delusional, greedy demoniacs who want to rule the world and enslave everyone else, but it is ultimately the violent conflict between the kingdom of the devil, hell on earth; clashing with the Kingdom of GOD, Heaven on earth. Kingdom against kingdom; which is why I am plainly calling every able bodied and willing saint to arms, because the days are now upon us when devils are seeking to do their evil deeds unrestrained and such evil deeds as should not be thought of let alone done.

https://www.blastthetrumpet.org/PublicLetters/AAAUpdatedPublicAlertsMattersofLifeandDeath/Updates053016/The%20Battle%20of%20Armageddon.pdf - like it or not there are only two ways to defeat delusional devils on earth who have cast off all restraint from doing exceedingly horrific wicked deeds. 1) Divine Intervention (flood the world, incinerate wicked cities, plagues, pestilences, storms, droughts, famines, etc.) and 2) RIGHTEOUS PEOPLE ARMED AGAINST THEM!

https://www.blastthetrumpet.org/PublicLetters/AAAUpdatedPublicAlertsMattersofLifeandDeath/Updates053016/the%20serpent%20seed.pdf - they were there in the past and they don't fit the evolutionary paradigm at all. The Holy Bible is true, not evil-u-shun!

https://www.google.com/search?q=artifacts+of+serpent+humanoids&rlz=1C1GIGM_enUS535US535&espv=2&biw=1066&bih=702&source=lnms&tbm=isch&sa=X&ei=4BtvVLKiFIz2yQSgqoHYBg&ved=0CAYQ_AUoAQ&dpr=1.5

https://www.google.com/search?q=artifacts+of+serpent+humanoids&rlz=1C1GIGM_enUS535US535&espv=2&biw=1066&bih=702&source=lnms&tbm=isch&sa=X&ei=4BtvVLKiFIz2yQSgqoHYBg&ved=0CAYQ_AUoAQ&dpr=1.5#tbm=isch&q=ancient+artifacts+of+reptilians

Understand I KNOW BEYOND ALL DOUBT the serpent seed, devils, Nephilim, Rephaim, are incarnate among us and HATE GOD and His People because we can see them and their deeds plainly and their REFUSAL to repent and honor the ONE TRUE GOD. Understand it is by direct Divine Revelation, that I know the inward spirits of these persons, the unclean spirits resemble all kinds of grotesque images like vampires, zombies, ghouls, goblins, imps, but also like what many refer to as dragons and dinosaurs, serpents, reptilians and not the glorified images of such but rather like snakes with arms and legs that could walk and talk. Now how "far-fetched" is the idea that some creatures that are presently referred to as "dinosaurs" could possibly communicate and were at war with or preyed upon humanity to a greater or lesser extent? How "far-fetched" is it when birds with brains smaller than a walnut can talk like us and even understand us?

Spiritually speaking, those Saved and Anointed by God appear like angels whereas those who aren't appear as unclean spirits, some just like what Hollywood depicts of them in movies. If you don't want to appear that way, you MUST REPENT AND BE BAPTIZED IN THE NAME OF OUR LORD JESUS CHRIST and then God cleans you and make you shine like the angels!

https://www.facebook.com/video.php?v=538809232843183 - animals are FAR more intelligent than most people give them credit for.

https://www.facebook.com/video.php?v=661737373940539

https://www.facebook.com/video.php?v=1226077770741575

https://www.facebook.com/video.php?v=10153177742633747

https://www.facebook.com/video.php?v=338754026282329

https://www.youtube.com/watch?v=2q6utsksoFY

http://www.sciencealert.com/watch-lyrebird-perfectly-imitates-laser-gun-sounds

https://www.youtube.com/watch?v=WA0tP-p7m40

https://www.facebook.com/ClimatologiaGeografica/videos/814007722025977/ - talk about a one-track mind :)

https://www.facebook.com/kq105/videos/10153002099497118/

https://www.facebook.com/sitecifras/videos/856119174428447/

https://www.facebook.com/PhoTeam/videos/546886905508850/ - soulish creatures benefit from bonding with us and likewise mankind benefits from them, by Divine Design.

https://youtu.be/d4EaWPIlNYY?t=2446

https://www.facebook.com/video.php?v=10155013935290019&set=vb.106854275018&type=2&theate r

https://www.youtube.com/watch?v=MHz-PALnprI

https://www.facebook.com/video.php?v=735470689851205

https://www.facebook.com/video.php?v=1073343389379517

https://www.facebook.com/video.php?v=851829108170842

https://www.google.com/webhp?sourceid=chrome-instant&ion=1&espv=2&ie=UTF8#q=elephants%20walk%20miles%20to%20honour%20their%20saviour

https://www.youtube.com/results?search_query=elephants+paint

https://www.youtube.com/watch?v=VjE0Kdfos4Y&app=desktop

https://www.facebook.com/video.php?v=771894312841591

https://www.google.com/webhp?sourceid=chromeinstant&rlz=1C1GIGM_enUS535US535&ion=1&espv=2&ie=UTF8#q=einstein+the+talking+texan+parrot&tbm=vid

https://www.facebook.com/video.php?v=10152054591023908

http://jezebel.com/ravens-are-terrifyingly-smart-and-form-complicated-rave-1567755951

https://www.google.com/webhp?sourceid=chromeinstant&rlz=1C1GIGM_enUS535US535&ion=1&espv=2&ie=UTF8#q=new+scientific+studies+on+ravens+show+cooperation+and+communication

https://www.youtube.com/results?search_query=talking+birds

https://www.youtube.com/results?search_query=bird+fishing+with+bread

https://www.youtube.com/watch?v=nMtn25zxT10

https://www.youtube.com/results?search_query=ravens+complex+problem+solving

https://www.facebook.com/Q1043NY/videos/10152560092601933/

https://youtu.be/MBm7c9poY8c?list=PLurUr4kT0BHVIRpZ-JZLP2C7hcUannUCm&t=893 - the soulishness of some creatures that science is now investigating has been uniquely published for millenniums in the Holy Bible.

http://www.sciencedaily.com/releases/2005/02/050223140605.htm - complex communications over thousands of miles and even identity awareness is one of the main reasons I am against sonar blasting technologies (I believe they are injuring the whales similar to destroying our own senses with over stimulation; like blinding light or loud blasts bursting ear drums or destroying sensitive nerves.)

https://www.google.com/webhp?sourceid=chromeinstant&rlz=1C1GIGM_enUS535US535&ion=1&espv=2&ie=UTF8#q=deep%20sea%20sonar%20pulsing%20injuring%20whales

Furthermore, identity awareness is a sign of extreme intelligence and whales should be off the menu or eaten as a last resort to prevent starvation.

https://www.google.com/webhp?sourceid=chrome-instant&ion=1&espv=2&ie=UTF8#q=how+do+dolphins+identify+themselves

There are many other reasons I am against the slaughter of dolphins and whales, not only the historic facts and current evidence as to their intelligence, socialization and intentional saving of human lives. I suppose just about all other fish can be harvested over them, but if you must to survive, they are not forbidden by God, I just prefer them over sharks or creatures that attack use because they are habitually gentle and have saved us

at times.
http://www.kosherquest.org/bookhtml/FISH.htm

In addition, I have had Divine Revelations that all Creation currently is designed to teach us good and bad behavior; to esteem the former and eschew the latter; even marine life imitates the struggle between good versus evil and the whales are definitely on the side of good; not so, the sharks.

https://www.youtube.com/watch?v=qket-48CGis

So with brains as enormous as in some of the "dinosaurs", it is likely they communicated with each other, some probably hunted in packs, and even hunted humans http://biblehub.com/genesis/3- 15.htm BEFORE YOU THINK OF LAUGHING AND MOCKING SUCH A STATEMENT CONSIDER OBSERVABLE FACTS EVEN TODAY! "The attacks began almost immediately, the Telegraph reports. A relentless lizard repeatedly came at a Swedish woman, who smacked it with her diving weight belt. It chewed at the lead belt while other divers threw rocks at its head, she said, all the while eyeing her bare feet." -

http://www.smithsonianmag.com/science-nature/the-most-infamous-komodo-dragon-attacks-of-thepast-10-years-5831048/?no-ist dragons/dinosaurs are STILL attacking people to this very day! we don't call them dragons (except for komodo dragons) like they did in past history today we call them alligators, crocodiles, hippos, rhinos, pythons, anacondas and other serpents and other such names instead of sea dragons we call them sharks, whales, octopus, squid and others such names. BUT NEVERTHELESS, many such creatures and more like giant jellyfish are deadly and objectively frightening if you were to see such up close and personal.

http://www.brainjet.com/world/14153/19-pictures-that-prove-you-definitely-have-thalassophobia-fearof-the-sea#slide/0

(not ALL dragons/dinosaurs mind you would prowl and attack humans if they got the chance but many most certainly would have and did), but I wonder if in the plethora of fossils we are perhaps not being told the FULL discoveries that have been made; like the suppression of giants all over the world, and all the humanoid oddities that have been found to date and made public, that perhaps they have found evidences of snakes with legs, and even reptilian like skulls near bones that didn't seem to belong; similar to all these misshapen humanoids -

https://www.youtube.com/watch?v=1zz8_MxcnzY):

https://www.google.com/search?q=artifacts+showing+dinosaurs+eating+people&rlz=1C1GIGM_enUS535US535&espv=2&biw=1066&bih=702&tbm=isch&tbo=u&source=univ&sa=X&ei=nQ1vVNaYD8uayATJ2ICoDw&ved=0CB0QsAQ&dpr=1.5

http://www.6000years.org/frame.php?page=dinosaurs_depictions

https://www.google.com/search?q=serpent+worship+throughout+history&rlz=1C1GIGM_enUS535US535&espv=2&biw=1066&bih=702&source=lnms&tbm=isch&sa=X&ei=flJvVOigN5SmyASSoGwAQ&ved=0CAcQ_AUoAg&dpr=1.5

http://www.godinanutshell.com/project/ancient-discovery/nephilim-the-evidence-history-nephilimdna-elongated-skulls-and-peru-part-1/

https://www.google.com/search?q=smithsonian+gatekeeping+knowledge+of+giants+and+other+facts+from+the+public&rlz=1C1GIGM_enUS535US535&oq=sm&aqs=chrome.0.69i59j0l2j69i61j69i57j69i60.3031j0j8&sourceid=chrome&es_sm=93&ie=UTF-8

https://www.google.com/search?q=ancient+serpent+worship&rlz=1C1GIGM_enUS535US535&espv=2&biw=1066&bih=702&tbm=isch&tbo=u&source=univ&sa=X&ei=ZVRvVJaFAYysyATyloKADA&ved=0CEMQs AQ

http://www.genesis6giants.com/ - giants have been documented throughout history (but they have no place in the evolutionary nonsense forced on the unsuspecting today) Such evidence is ignored; even though the Biblical Creation Account tells us exactly where they came from.

https://www.google.com/search?q=human+sacrifice+to+serpents&rlz=1C1GIGM_enUS535US535&espv=2&biw=1066&bih=702&source=lnms&tbm=isch&sa=X&ei=3lRvVP7uKdeyyATVh4DQCA&ved=0CAcQ_AUoAg

Which rises throughout history to modern wars, genocides, abortions, and the shedding of innocent blood globally.

https://www.ccel.org/bible/phillips/CPn27Revelation2.htm the shedding of innocent blood portends doom for every wicked nation on earth condoning such practices.

http://biblehub.com/proverbs/6-17.htm Warnings against Foolishness ...16There are six things which the LORD hates, Yes, seven which are an abomination to Him: 17Haughty eyes, a lying tongue, And hands that shed innocent blood,

18A heart that devises wicked plans, Feet that run rapidly to evil,… All nations are called to repent immediately of this great evil and return in true repentance to the ONE TRUE GOD, the LORD our GOD, our CREATOR and the SAVIOR of all who put their trust in HIM!!!!!!!!

https://www.blastthetrumpet.org/PublicLetters/AAAUpdatedPublicAlertsMattersofLifeandDeath/Updates053016/Observable%20Divine%20Judgments.pdf - The Holy Bible is the foundation for public education in our nations (USA AND WESTERN CHRISTIANIZED NATIONS) and needs to REMAIN AS SUCH, those who have plotted and planned for its removal do not have our nations interest at heart!

https://www.blastthetrumpet.org/PublicLetters/AAAUpdatedPublicAlertsMattersofLifeandDeath/Updates053016/Americans%20and%20Citizens%20of%20the%20World%20FREEDOM%20and%20LIFE%20Comes%20from%20GOD.pdf - Our freedoms come from a moral and righteous citizenry UNDER GOD ALMIGHTY, BIBLICALLY BASED!

https://www.blastthetrumpet.org/PublicLetters/AAAUpdatedPublicAlertsMattersofLifeandDeath/Updates053016/Choose%20LIFE%20Avoid%20Death%20and%20Destruction.pdf - America MUST choose Jesus Christ, to depart from Him spells doom to our nation, its cities, etc.

The written history of man shows that, contrary to the fictional constructs lying to mass millions by claiming that "dinosaurs" went extinct "millions of years ago", ones in recent history that have now gone extinct, became such with GOOD REASON; namely that they attacked and ate people! (like the ones that are still with us do):

https://www.google.com/search?client=opera&q=man+eating+crocodiles+and+alligators&sourceid=opera&ie=UTF-8&oe=UTF-8

https://www.google.com/search?q=man+eating+sharks&rlz=1C1GIGM_enUS535US535&oq=man+eating+sharks&aqs=chrome..69i57&sourceid=chrome&es_sm=93&ie=UTF-8

https://www.google.com/search?q=man+eating+komodo+dragons&rlz=1C1GIGM_enUS535US535&oq=man+eating+komodo+dragons&aqs=chrome..69i57&sourceid=chrome&es_sm=93&ie=UTF-8

https://www.google.com/webhp?sourceid=chromeinstant&rlz=1C1GIGM_enUS535US535&ion=1&espv=2&ie=UTF-8#q=man%20eating%20snakes

https://www.google.com/webhp?sourceid=chromeinstant&rlz=1C1GIGM_enUS535US535&ion=1&espv=2&ie=UTF-8#q=man%20eating%20hippopotamus

My point here is that even though it is obvious the tree of the knowledge of evil and the talking serpent are clearly metaphors for the devil, wicked deceptive spirit, that it is NOT impossible that in the past there were serpents, dragons, dinosaurs with arms and legs that could have communicated and with the presence of evil spirits even mimicked human speech. It is certain that GOD has such capability.

http://biblehub.com/2_peter/2-16.htm and http://biblehub.com/matthew/19-26.htm

But I point this out, that while atheists and evolutionists believe in PROVEN IMPOSSIBILITIES with zeal and religious fervor; even what they ridicule Biblical Creationists about is scientifically feasible; by what we plainly observe. The serpent seed, over time, may have genetically altered their outward appearance; or the

Nephilim, Rephaim and such humanoid hybrids may have even interbred into even more human-like features; but their deeds always manifest who they really are.

https://www.blastthetrumpet.org/PublicLetters/AAAUpdatedPublicAlertsMattersofLifeandDeath/Updates053016/satanism%20is%20a%20danger%20to%20everyone%20even%20those%20who%20practice%20it.pdf - devil worshipers are not in their right minds.

https://www.blastthetrumpet.org/PublicLetters/AAAUpdatedPublicAlertsMattersofLifeandDeath/Updates053016/satanil%20v%20GOD%20part%201.pdf - no one should worship the destroyer unless they want to be destroyed with him in the lake of fire. (part 1)

https://www.blastthetrumpet.org/PublicLetters/AAAUpdatedPublicAlertsMattersofLifeandDeath/Updates053016/satanil%20v%20GOD%20part%202.pdf (part 2)

https://www.blastthetrumpet.org/PublicLetters/AAAUpdatedPublicAlertsMattersofLifeandDeath/Updates053016/satanil%20v%20GOD%20part%203.pdf - (part 3) the devil, satanil, is a deceiver and so all worldviews that do not point to Jesus Christ as the One True God are deceptive, and of the devil. You MUST REPENT and acknowledge the ONE TRUE GOD, Jesus Christ, He is the ONLY ONE WHO CAN SAVE YOU and make sure you go to Heaven! (Acts 4:12, Acts 2:38,39 Acts 1:7-11) Understand, these are just my personal theories or thoughts, about the Serpent Seed, talking snake, based on my own reading comprehension and what has been documented about the ancient past. I think its not just spiritually identifying people as "serpents" based on their spiritual behavior, but also the history of the devil and his children in this fallen world. Matt 3:7-9

The entire content of the HOLY BIBLE is historically and scientifically accurate; not just concerning our past and present but has a 100% track record concerning prophecies, many thousands of them. It is time all those mocking and laughing really compared what they THINK they KNOW to the OBSERVABLE EVIDENCE.

https://www.facebook.com/photo.php?fbid=1009149802463123&set=a.345811072130336.86973.100001043398117&type=1&theater -

Creationists are not denying WHAT WE OBSERVE, but evolutionist take what we observe and then apply their fantasy that every living thing came from a single celled organism that "evolved" (i.e. ADDED genetic information to itself countless trillions of times to become more complex organisms, mass millions of variations resulting in all species that have ever existed (even with their fantasies of hundreds of millions to billions of years, such fiction is NOT science); it isn't even science fiction, because it darwinian macro-evolution is simply not true, not even possible even though it is still forced on innocent children today.) I reiterate READ AND LOOK AT EVERY CITATION if you do not KNOW our ETERNAL CREATOR personally as yet!!!!!!!! Help people not to present themselves as https://www.blastthetrumpet.org/PublicLetters/AAAUpdatedPublicAlertsMattersofLifeandDeath/Updates053016/Mentally%20Challenged.pdf by teaching the Truth!

https://www.blastthetrumpet.org/PublicLetters/AAAUpdatedPublicAlertsMattersofLifeandDeath/Updates053016/Evidence%20of%20GOD.pdf and

https://www.blastthetrumpet.org/PublicLetters/AAAUpdatedPublicAlertsMattersofLifeandDeath/Updates053016/Teach%20the%20Children%20Truth.pdf - Teach the Children TRUTH! BE ROCK SOLID CERTAIN YOU ARE DOING AS GOD COMMANDS, TEACH THEM OUT OF THE HOLY BIBLE!

Far too many people today THINK the contents of the Holy Bible have no way to be "scientifically" tested in repeatable procedures BUT THAT IS A MOST UNTRUE ASSUMPTION! When you apply the so called "falsification principle" to the Holy Bible, one finds instead over and over again that what it says is fundamentally TRUE; and supported by what WE ACCURATELY OBSERVE (science) throughout!

https://www.youtube.com/watch?v=ZlQllBxLRa4

https://www.youtube.com/watch?v=lu_Hp3dq7KY

https://www.youtube.com/watch?v=5Dlsw9wsN60

https://www.facebook.com/VideoNatGeo/videos/1229941810368614/

https://www.facebook.com/Upworthy/videos/1245827138791475/

http://store.livingwaters.com/index.php?page=shop.product_details&flypage=flypage.tpl&product_id=29&category_id=7&option=com_virtuemart&Itemid=368&lang=en

Actually, observable evidence supporting the Biblical Account is so overwhelming, that people have to force themselves to ignore it in order to remain in other worldviews (especially in the madness of atheism and evolutionary imaginary constructs) -

https://www.blastthetrumpet.org/PublicLetters/AAAUpdatedPublicAlertsMattersofLifeandDeath/Updates053016/Evolution%20and%20Atheism%20Intertwined%20Cults%20of%20the%20Insane.pdf).

http://blogs.answersingenesis.org/blogs/ken-ham/2014/11/14/scientists-say-atheists-might-not-exist/

https://www.blastthetrumpet.org/PublicLetters/AAAUpdatedPublicAlertsMattersofLifeandDeath/Updates053016/Acknowledging%20the%20Eternal%20Creator%20Takes%20No%20Faith%20It%20is%20Scientific%20Fact.pdf - Acknowledging our Eternal Creator only takes accurate information and a rational mind.

It is this ubiquitous observation that our Creator made us all inherently "religious" (ever noticed how those who claim to be irreligious, zealously defend that claim?); ever notice how what the atheists believe (pure fantasy and delusions, requires more blind faith than all the major "religious" worldviews combined); thus even the so called "irreligious" have a set of beliefs (the modern evolutionary paradigm) that I have just plainly shown is entirely religious in nature; in that they substitute themselves and so called scientists as their deities arguing infallibility on topics ALREADY DISPROVEN MANY TIMES OVER; like macro-evolution, abiogenesis and claiming the universe itself came from nothing, order from chaos, more complexity from less complexity, life from non-life, intelligence from ignorance, and all the other nonsense and rubbish these people believe because they cannot face the reality of our Eternal Creator. It is for those reasons and observations of humanity all over the world that I have written such statements regarding the impossibility of "separation of church and state". It is impossible to separate the authorities (governments to govern "the state") as it is impossible to separate our Eternal Creator from His Creation. (Matt 28:18-20)

https://www.blastthetrumpet.org/PublicLetters/AAAUpdatedPublicAlertsMattersofLifeandDeath/Updates053016/Original%20Intent%20of%20the%20Religious%20Clause%20of%20the%20First%20Amendment.pdf FREEDOM OF RELGION AND TO SPEAK (PREACH THE GOSPEL LOUDLY AND PROUDLY) and the second one (right to bear arms was to make certain no one took our first amendment away from us) and

https://www.blastthetrumpet.org/PublicLetters/AAAUpdatedPublicAlertsMattersofLifeandDeath/Some%20Quotes%20and%20Thought1.pdf - Some thought about FREEDOM! It's most definitely does not occur under anything other than Jesus Christ and His Righteous Commandments.

AS WE CAN EASILY SEE HERE IN AMERICA, the removal of the HOLY BIBLE and TRUTH from public education is resulting in MASSIVE IGNORANCE, BRAINWASHING and DECEPTIONS! (even to the point of allowing the vilest of all worldviews to creep in and take hold, satanism and islam).

https://www.blastthetrumpet.org/PublicLetters/AAAUpdatedPublicAlertsMattersofLifeandDeath/Updates053016/THE%20PLANNED%20DEMORALIZATION%20OF%20AMERICA.pdf - planned demoralization is a tragedy for any nation, corruption leads to its downfall.

https://www.blastthetrumpet.org/PublicLetters/AAAUpdatedPublicAlertsMattersofLifeandDeath/Updates053016/satanism%20is%20a%20danger%20to%20everyone%20even%20those%20who%20practice%20it.pdf - Project paperclip brought eugenic fascism here to America and shortly after WWII our nation began the removal of the

Holy Bible and creation science and into indoctrinating with evil-u-shun. It was an ATTACK ON AMERICA CHRISTIAN values that is more politically motivated than anything that has to do with science! IT WAS A DELIBARATE ATTEMPT TO UNDERMINE THE HOLY BIBLE and corruption was behind it!

https://www.blastthetrumpet.org/PublicLetters/AAAUpdatedPublicAlertsMattersofLifeandDeath/Updates053016/America%20is%20Under%20Attack.pdf - in just a few generations we now have opened our doors to openly antichristian worldviews and criminal organizations like islam! EVERY ATTEMPT TO UNDERMINE THE HOLY BIBLE IS AN ATTACK ON OUR NATION!

https://www.blastthetrumpet.org/PublicLetters/AAAUpdatedPublicAlertsMattersofLifeandDeath/Updates053016/satanil%20v%20GOD%20part%201.pdf - brainwashing kids with the same kind of evil crap that led to the holocaust!

https://archive.org/details/DarwinsDeadlyLegacy

https://www.blastthetrumpet.org/PublicLetters/AAAUpdatedPublicAlertsMattersofLifeandDeath/Updates053016/satanil%20v%20GOD%20part%202.pdf - people actually think its science, when it was an agenda of people "who wanted WORKERS (slaves) to do his bidding, rather than thinkers." Rothschilds and other wealthy people took over the board of education and ripped the Holy Bible away from our public schools to brainwash children with their evil-u-shun!

https://www.blastthetrumpet.org/PublicLetters/AAAUpdatedPublicAlertsMattersofLifeandDeath/Updates053016/satanil%20v%20GOD%20part%203.pdf - when you do away with the TRUTH that made our nation Great, GOD- JESUS CHRIST, all you get in His Place are LIES! corruption! this is the real reason our nation is SUFFERING more and more; WE MUST RETURN TO GOD!

https://www.blastthetrumpet.org/PublicLetters/AAAUpdatedPublicAlertsMattersofLifeandDeath/Updates053016/PROPAGANDA.pdf - we have to make sure the fourth estate keeps us on the path of TRUTH, rather than that it's been taken over by wealthy people who tell us what they want us to hear!

https://www.blastthetrumpet.org/PublicLetters/AAAUpdatedPublicAlertsMattersofLifeandDeath/Updates053016/islam%20is%20Evil%20part1.pdf - otherwise we become susceptible to enemy infiltration and our nation suffers.

https://www.blastthetrumpet.org/PublicLetters/AAAUpdatedPublicAlertsMattersofLifeandDeath/Updates053016/islam%20is%20Evil%20part%202.pdf - Deuteronomy 28 one of the consequences of departing from God is enemies come into our borders as is happening right now!

https://www.blastthetrumpet.org/PublicLetters/AAAUpdatedPublicAlertsMattersofLifeandDeath/Updates053016/islam%20is%20Evil%20part%203.pdf - more and more godless people who do not KNOW our Lord and Savior JESUS CHRIST!

https://www.blastthetrumpet.org/PublicLetters/AAAUpdatedPublicAlertsMattersofLifeandDeath/Updates053016/islam%20is%20organized%20crime.pdf - yes! I believe as soon as we started brainwashing children with evil-u-shun and away from God, they stopped being an informed American. Not understanding it's our CHRISTIAN HERITAGE that made our nation Great! And so, we are suffering ever since. Stop brainwashing children with this impossible nonsense! PREACH THE HOLY BIBLE! IT IS VERIFIABLY TRUE!

https://www.blastthetrumpet.org/PublicLetters/AAAUpdatedPublicAlertsMattersofLifeandDeath/Updates053016/Warning%20America%20Has%20Been%20Infiltrated%20and%20Invaded.pdf - We have got to stop making Saudi Arabia, islamic nation, rich by buying oil from them! islam is nothing like our freedoms and our cultural heritage AT ALL!

https://www.blastthetrumpet.org/PublicLetters/AAAUpdatedPublicAlertsMattersofLifeandDeath/Updates053016/Conspiracy%20Theories%20or%20Plain%20Truth.pdf - We need to reverse everything they have done to corrupt our nation and remove our Godly Heritage NOW! PUT THE HOLY BIBLE BACK IN OUR SCHOOLS AND NEVER, EVER LET IT DEPART FROM THEM!

https://www.blastthetrumpet.org/PublicLetters/AAAUpdatedPublicAlertsMattersofLifeandDeath/Updates053016/Most%20Wanted%20Criminals%20part%201.pdf - World's most wanted criminals. Part 1

https://www.blastthetrumpet.org/PublicLetters/AAAUpdatedPublicAlertsMattersofLifeandDeath/Updates053016/Most%20Wanted%20Criminals%20part%202.pdf - most wanted part 2

https://www.blastthetrumpet.org/PublicLetters/AAAUpdatedPublicAlertsMattersofLifeandDeath/Updates053016/Most%20Wanted%20Criminals%20part%203.pdf - most wanted part 3

https://www.biblegateway.com/passage/?search=Deuteronomy%2028&version=AMP - this chapter deals with rewards and consequences. These rewards and consequences can be seen throughout history to this very day! If the people of any nation became wicked, they suffered the consequences; or if they turned to our Creator and followed His Instructions to Live By, they were blessed and prospered accordingly.

Read my citations on: The Sufferings of Christ, The Ten Commandment (Part 1-3), End Times Verse by Verse, and Observable Divine Judgments. (cited redundantly in this book)

But the most self-evident application of the Scriptures one can perform is by following the Instructions (just like one follows a recipe, they MUST follow it precisely to achieve the results); whereby those who do so will personally encounter our Creator. Very precise instructions are given on how to approach and KNOW the One True GOD, who will thereafter personally teach all who follow those steps to KNOWING Him and beginning to partake of His Glorious Kingdom HERE and NOW and FOREVER!!!

https://www.blastthetrumpet.org/PublicLetters/AAAUpdatedPublicAlertsMattersofLifeandDeath/Updates053016/Religion%20in%20the%20Name%20of%20Science.pdf - while hammering on about teaching Creation WHICH IS TRUE, you ended up teaching your cultic beliefs "darwinism" THAT ISN'T TRUE!

YOUR BELIEF IS RELIGIOUS more particularly a form of idolatry in a cult originally founded by Darwin! There is NOTHING scientific, NO empirical evidence to support evolution NOTHING! FACE IT! If evolution were true INTERMEDIARY SPECIES between the dogs (mammals) and whales would STILL BE ALIVE, likewise between APE-LIKE creatures and HUMANS! and all the OTHER IMAGINATIONS - horse-like and giraffes! but we have millions of each of the supposed results of evolution, and we still have species that SUPPOSEDLY are like their common ancestors with NOTHING in between them. The fossil record shows diversity of species many of which have become extinct; but the BILLIONS of intermediaries are NO WHERE TO BE FOUND!

Contrary to evolution, the Biblical Account DOES HAVE OBSERVABLE EVIDENCE, DOES HAVE SCIENTIFIC SUPPORT THROUGHOUT! Creation is the REALITY in which we exist (hands down! no contest whatsoever between Biblical Creation and imaginary evolution!) - evolutionists instead of throwing up ad hominems and their usual nonsense need to consider the FACTS cited in these two presentations

https://www.youtube.com/watch?v=GjvuwneORrE and

https://www.youtube.com/watch? v=lktmmd7YnD8

You see, Biblical Creationists DO NOT DISMISS THE RECORDED HISTORY of mankind that tells us PLAINLY about Creation and is FACTUALLY supported by what we observe, but evolutionists SWEEP history aside COMPLETELY and assign their own fairy tales to what we observe.

https://www.blastthetrumpet.org/PublicLetters/AAAUpdatedPublicAlertsMattersofLifeandDeath/Updates053016/The%20Modern%20Evolutionary%20Paradigm%20is%20NOT%20SCIENCE%20part%201.pdf - the evolutionary paradigm taught today IS NOT SCIENCE, IT DOESN'T EVEN RISE TO THE MERIT OF SCIENCE FICTION! part 1 of 5

https://www.blastthetrumpet.org/PublicLetters/AAAUpdatedPublicAlertsMattersofLifeandDeath/Updates053016/The%20Modern%20Evolutionary%20Paradigm%20is%20NOT%20SCIENCE%20part%202.pdf - part 2 of 5

https://www.blastthetrumpet.org/PublicLetters/AAAUpdatedPublicAlertsMattersofLifeandDeath/Updates053016/The%20Modern%20Evolutionary%20Paradigm%20is%20NOT%20SCIENCE%20part%203.pdf - evolution is IMPOSSIBLE. part 3 of 5

https://www.blastthetrumpet.org/PublicLetters/AAAUpdatedPublicAlertsMattersofLifeandDeath/Updates053016/The%20Modern%20Evolutionary%20Paradigm%20is%20NOT%20SCIENCE%20part%204.pdf - evolution is built on lies and more lies, completely fabricated! part 4 of 5

https://www.blastthetrumpet.org/PublicLetters/AAAUpdatedPublicAlertsMattersofLifeandDeath/Updates053016/The%20Modern%20Evolutionary%20Paradigm%20is%20NOT%20SCIENCE%20part%205.pdf - the Holy Bible is factually true with well verified evidence. The theory of evolution is soundly disproven. Part 5 of 5

PART OF THAT MOST SIGNIFICANT HISTORY BEING IGNORED IS HOW THE CREATION ACCOUNT IS TOLD IN THE STARS AND EVEN HOW AND WHEN OUR CREATOR WOULD COME AND SAVE MANKIND!

https://www.youtube.com/watch?v=EUQEMqF5dL8

https://www.blastthetrumpet.org/PublicLetters/AAAUpdatedPublicAlertsMattersofLifeandDeath/Updates053016/GOD%20Part%201.pdf - God is our Eternal Creator (part 1) and

https://www.blastthetrumpet.org/PublicLetters/AAAUpdatedPublicAlertsMattersofLifeandDeath/Updates053016/GOD%20Part%202.pdf - Jesus Christ is just what the Holy Bible says He is - the One TRUE GOD! (part 2)

GET THIS CLEAR; VERY, VERY CLEAR; LET IT SINK IN! SCIENCE OVERWHEMINGLY SUPPORTS THE BIBLICAL ACCOUNT; SCIENCE DOES NOT SUPPORT the entire evolutionary paradigm currently being forced upon innocent children in coerced public indoctrination and reinforced through deceptive media programming on television; from cosmic evolution on through to biological macro evolution! **LET ME REPHRASE THIS; SCIENCE SUPPORTS, CONFIRMS, PROVES AS VALID THE BIBLICAL ACCOUNT (IS WHAT WE ACTUALLY OBSERVE); SHOWING BEYOND ALL REASONABLE DOUBT, THAT IT IS MOST DEFINITELY THE CREATIONISTS THAT ARE THE ONES WHO BASE THEIR FINDINGS, CONCLUSIONS AND WORLDVIEW**

ON REASON AND SCIENCE; NOT ANY of those who cling to and espouse evil-u-shun!

https://www.youtube.com/watch?v=t8H35GDLNuw&feature=share

Such nonsense being forced on unsuspecting children during their development years in coerced public indoctrination is leaving people today by the mass millions as having the appearance of being

https://www.blastthetrumpet.org/PublicLetters/AAAUpdatedPublicAlertsMattersofLifeandDeath/Updates053016/Mentally%20Challenged.pdf - we need THINKERS who KNOW GOD!

GRASP THIS CONCEPT AND NEVER FORGET IT!!!!!!!! INFORMATION or INTELLIGENCE is ABSOLUTELY necessary for life!!!!!!!! In other words, ALL LIFE is EVIDENCE of our ETERNAL, EVER-LIVING, CREATOR. I have listened to naturalists describe chemicals, elements as the "hardware" and DNA as the "software" (INFORMATION) necessary for and observed in ALL living organisms. TO HAVE BOTH HARDWARE AND SOFTWARE IN ANY SYSTEM REQUIRES AN ENGINEER; THAT ENGINEER has made

HIMSELF CLEARLY KNOWN. He is the ONE TRUE GOD, OUR LIVING LORD JESUS CHRIST; WHO IS EVER PRESENT TO SAVE ALL WHO CALL UPON HIM!!!!!!!! All persons OMITTING INFORMATION/INTELLIGENCE as ABSOLUTELY NECESSARY FOR LIFE are in

https://www.blastthetrumpet.org/PublicLetters/AAAUpdatedPublicAlertsMattersofLifeandDeath/Updates053016/Denial%20of%20REALITY.pdf - denial of reality.

Everyone who knows the Lord, please ask evolutionists and atheists for me:

https://www.blastthetrumpet.org/PublicLetters/AAAUpdatedPublicAlertsMattersofLifeandDeath/Updates053016/Please%20Dont%20Choose%20to%20be%20Stupid.pdf - it's one thing for someone who was deceived by brainwashing to be ignorant due to that brainwashing but it's another when they have the truth right in front of them but go on in willful ignorance. That kind of willful ignorance in denial of the ONE TRUE GOD WHO CAN BE KNOWN, is called "stupid on purpose". Don't chose to be one of them.

People who have accurately studied world history know that at any given time; not just in the past but even today, you can pick up a modern manufactured artifact and hand it to billions of people all over the world and many will NOT know what it is or what it is for! The

MANUFACTURER provides INSTRUCTIONS for its use, telling those who READ what it is exactly. That is the SAME error people are making (calling themselves scientists) when they look at Creation and IMAGINE what it means, rather than reading the Instructions from our Creator telling them so plainly!

https://www.blastthetrumpet.org/PublicLetters/AAAUpdatedPublicAlertsMattersofLifeandDeath/Updates053016/The%20cult%20of%20modern%20scientism%20versus%20the%20Science%20of%20Biblical%20Creation.pdf - The proponents of the impossible nonsense of evolutionary thought, ARE PUTTING THEIR CULTIC BELIEFS in the name of science! They are doing exactly what they fault creation scientists for doing in our class rooms! THERE WAS NOTHING WRONG WITH CREATION SCIENCE AS BEING TAUGHT OUR CHILDREN UNTIL THEY WRONGLY DECIDED THAT RELIGION HAD NO PLACE IN OUR SCHOOLS. That one decision ripped public education away from how it should be CHILDREN TAUGHT TRUTH IN OUR SCHOOLS, regardless if it's considered religious or not! **THE BIBILICAL GOD CREATED CREATION - THERE IS NOTHING WRONG WITH TELLING CHILDREN THE TRUTH!**

https://www.facebook.com/photo/?fbid=1408899032522385&set=a.115635768515391 - Meditate on the Holy

Bible, it has the Words of God and leads us into all blessings!

https://www.facebook.com/photo/?fbid=2385941711484774&set=a.115635768515391 - your ancestors were NOT an evolved ape-like creature or anything else like they're teaching you! **YOU ARE A DIVINE CREATION WITH DIVINE PURPOSE!** I list MANY examples of this fact in my book, The Holy Bible is Provably and Verifiably Divinely Inspired.

https://www.facebook.com/photo/?fbid=2322419994503613&set=a.115635768515391 - **if your presently someone in denial of GOD ALMIGHTY, Jesus Christ, then you need to remove all the garbage in your head you presently mistake for knowledge and LEARN FROM HIM! REPENT AND BE BAPTIZED IN HIS NAME, HE WILL GIVE YOU HIS HOLY SPIRIT and teach you only TRUTH!**

https://www.facebook.com/photo/?fbid=4890242114388042&set=a.115635768515391 - **The Holy Bible is the most studied, most thoroughly verified, and most accurate REFERENCE BOOK IN ALL THE WORLD - BY FAR!** Never let anyone disdain it! Any doing so are demonstrating they are completely lacking a proper education!

https://www.facebook.com/photo?fbid=6382847548460817&set=a.115635768515391 - **Jesus Christ is the TRUTH you MUST KNOW in order to not be deceived by all the lies in this world!**

Symbiotic Relationships

Reason 6 of why evolution is not just false but impossible are symbiotic relationships. All sexually reproducing organisms need both males and females to reproduce or they wouldn't exist. So slight changes over massive amounts of time, in no way accounts for the fact that mass millions of times over there had to have been MASSIVE changes for any of the many tightly woven symbiotic relationships to exist at all. On the one hand, evolutionists will claim that MASSIVE AMOUNTS OF INFORMATION AND MATERIAL can be explained by tiny amounts added over time and yet on the other, point to another life form and say it's losing information, structure and function, the opposite of their common claims:

https://www.livescience.com/first-non-breathing-animal.html - an animal that doesn't breathe and the supposed explanation is that it has lost genetic information over time (DEVOLVED; not evolved!)

"Dictionary.com

The opposite of evolution is devolution. It means "the evolutionary decline or loss of a function, characteristic, or structure in an organism or a species."

The fossil record shows that huge variety of live forms all once existed on this planet. It shows contrary to evolution LARGER live forms of virtually every living thing in the past. CREATION WAS NEW IT WAS AS FULL OF LIFE AS THE SCRIPTURES MAKE CLEAR. Then the flood happened and a wide range of extinction events occurred and have been occurring ever since. It is all just as our Creator tells it!

So, evolutionists want to argue that entropy, extinction events and evidence of devolution in front of their eyes, isn't proof that macro-evolution is false; even though it is. Unless there were massive leaps that can't ever be seen, ever be observed anywhere in history or the presumed to be "prehistoric" fossils, evolution is just plain false and as other laws of science and facts of history show: impossible.

Mankind has learned that specific molecules make up DNA and that depending on how these chemicals are arranged results in the various genomes, but I have yet to read or see anyone demonstrate, how common chemicals, sugar and proteins, result in the huge variety of life forms on earth. People talk about the hardware, essentially being hydrocarbons (hydrogen and carbon molecules) and DNA being the software that manipulates substances found in earth and water to form living organisms. How? How do chemicals, "nucleotides and bases": * Adenine (A)* Cytosine (C)* Guanine (G)* Thymine (T) tell the elements found in water and the earth to make various complex living creatures?

Do you understand my question(s), How precisely, are nucleotides and bases, telling physical matter, other chemicals, sugars, proteins, to form all the many various cells, the TRILLIONS of cells, necessary for every organ and every system of our bodies to all work together to form our functioning bodies, specifically in the form of MANKIND or any other KIND of creature? Is life really only defined by DNA or is there something more at work that mankind has yet to discover or is unwilling to identify? Just HOW can the exact same 4 molecules be rearranged to form a fish with fins and scales or a mammal with claws and fur or a bird with beaks and feathers? If physical matter is all there is to life, then why do we live at all and why do we die? If DNA, earth, water, chemicals, molecules, etc. are all that's necessary, then why are some babies born dead? (God Almighty tells us in the Holy Bible that He breathes life into us and that breath of God is called our spirit. He says we are more than just physical matter, more than just DNA, but that He Himself imbues our bodies with spirits that brings our flesh to life. He further tells us that He forms us in the womb.) So based on what our Eternal Creator tells us, our physical forms (what KIND of creature) are being determined by the seed, genetic makeup in each of us, but life itself is His Gift to us, His Own Breath, Spirit given to each of us. God in all, through all and above all; such that what is seen is not made up of what's visible but by that which is unseen, our Eternal Creator.

There are so many questions scientists still don't know the answers to, but one thing is certain the theory of evolution is false and impossible based on truthful knowledge; actual science. Some have experimented enough to find that certain parts of various genomes relate to eyes, and other body parts of various organisms, but no one (at least in all my research) knows how these four bases, these molecules, convey information so specific and vast as to control and coordinate trillions of cells in our bodies lifelong, to tell some billions of cells to form our brain, and others to form our heart and so on. Those who are growing skin cells in labs are only doing so by taking the already existing DNA and programming certain kinds of stem cells, (fibroblasts or embryonic stem cells) with that DNA, but to the best of my knowledge, no one has ever been able to recreate either a living cell, or artificial DNA that takes substances from water and earth, chemicals and molecules, and recreates a living creature thereby. In other words, I haven't been able to find even one scientist in all the world who can explain or demonstrate their own imaginary processes to all the rest of us. And if they could, it would be after intense amounts of study, research, trial and errors; which means, there's no way life arose from random, undirected, simple so-called "natural" processes EVER! because none of those posing themselves as geniuses or experts worldwide; despite diligently trying, can demonstrate, even with concerted and very best efforts throughout their lifetimes, how any of these basic building blocks of life came into existence! The truth is even the simplest life form is more complex than anything mankind has ever built themselves. How can I say that? because no one has ever been able to **recreate** (even after spending a lifetime of intensely observing what God, Jesus Christ, has already created and made) those base nucleotides from nothing but chemicals, even though they claim to understand those molecules, no one has been able to recreate the proteins necessary for even a single cell, let alone taken nothing but those chemicals they've identified and created even the most basic bacteria.

It appears to me that there's still so much about DNA, so much information and knowledge still hidden and yet to be discovered, that overwhelmingly points to Intelligent Design, that it becomes shameful for anyone not to acknowledge that fact. If even the greatest minds on earth all combined cannot recreate even the simplest life form, even with intense amounts of examination and effort, then to go on imagining life came into existence by abiogenesis, spontaneous generation, is just evidence that such persons are DESPERATE to deny the obvious. The universe and life exists because of our Eternal Creator, no ifs, ands, or buts. Even if one day mankind was to succeed in recreating a life form, it would still only prove Intelligent Design and overwhelmingly so. (if it takes the concerted effort of billions of presumably intelligent persons to recreate even the simplest life form, that means life exists by INTELLIGENCE; not chaotic, random so-called "natural" processes). So, abiogenesis and evolution are false and obviously impossible. The Holy Bible is telling us all the absolute Truth.

All recent significant changes (outside of the common variations within each kind of creature) in living organisms today have been due to intentional genetic modification. What some classify as a new species, would be as wrong as saying different skin colors indicate different species of humans. In other words, God clearly tells us that if any creature reproduces, it only reproduces like KIND of creature, so among all sexually reproducing organisms, only male and females of the same KIND of creature can mate and have offspring. (whereas speciation, makes separations within the same KINDS of creatures; which is why I reject not just the false theory of evolution, but all its related lingo, imaginary terms, words, and outright lies). I don't reject biology, micro-biology, or any of the sciences, I reject any lies being taught in the name of science, like the theory of evolution (and the entire evolutionary paradigm being taught today of how the universe came from nothing, or an imaginary "singularity" that exploded and created everything). NO! the universe didn't come from anything imaginary and NO! atoms, elements, stars and planets did not assemble themselves, etc. etc. etc. like any of these God-denying, deceived and deluded persons have proposed. Their story is completely untrue and impossible by everything we know and I say so plainly enough, citing the reasons why.

https://www.facebook.com/photo/?fbid=6382847548460817&set=a.115635768515391 and

https://www.facebook.com/photo.php?fbid=2385941711484774&set=pb.100002069048072.-2207520000&type=3 and

https://www.facebook.com/photo.php?fbid=1380774058668216&set=pb.100002069048072.-2207520000&type=3 and

https://www.facebook.com/photo.php?fbid=1316005115145111&set=pb.100002069048072.-2207520000&type=3 and https://www.facebook.com/photo.php?fbid=977229375689355&set=pb.100002069048072.-2207520000&type=3

When diseases or insects or anything survives anti-biotics, herbicides, pesticides or other adverse circumstances, all that is evidence of is that life has been designed to be resilient, just like our own circulation and immune systems resist infections and heal our sick or injured bodies. It doesn't mean anything is "evolving" to survive adversity, even if that same adversity kills others of its kind. Any mutations, all defects of any living creature, including mankind are still defects; in no way is evolutionary progress toward anything superior or more complex, any so-called "natural" causes of damage or alteration to DNA have only been harmful, never beneficial. A person might be born without arms and legs or perhaps an additional limb (that doesn't function because that limb needs the entire complex neurological and skeletal-muscular system to cause it to do so) So, DNA can be lost, damaged or a sequence copied badly, but it NEVER evolves or morphs into an entirely different and new functioning kind of creature, NEVER! In other words, all the processes we do know about how DNA is passed on between generations and to cells within the various kinds of life forms, creatures, shows us plainly that the theory of evolution is false and impossible. EVERYTHING observed throughout history, that alters DNA (apart from intentional genetic modification), is detrimental; harmful environmental causes leading to defects, loss of genetic information due to entropy; eventually leading to extinction. And yes, I've read about epigenetics and viruses inserting and altering some extra nucleotides or otherwise interfering with DNA transferred between cells and generations of the same kind of creatures. STILL HARMFUL, NOT beneficial.

When evolutionists, including Dawkins, are asked to give us even one example of a beneficial mutation from ANY of their imagined processes, from all recorded history, or anything observed anywhere in this world, there is nothing but silence. All alterations to DNA in the real world are harmful. And that's due to the fact that even though we're classifying genomes, there's still so much information in them that no one fully understands; so, a mere virus cannot cause a land-dwelling mammal to become a whale, not in millions or billions of imagined years. NONE of their proposed processes to date justify macro-evolution at all. NONE. And that's because DNA in every living organism is designed just as God tells us in the scriptures to reproduce ONLY LIKE KIND unto itself! Some creatures might get an extra arm or leg, or some other body part, but that was due to an err in the process of development, they don't get something completely different from their existing DNA, like a mammal suddenly gaining gills, or a beak or a tortoise shell. (And yes, I've heard all the whining from evolutionists claiming, that's not the way it works; even though at the same time they're claiming all life on earth "evolved" from a single common ancestor: all plants and animals, reptiles and mammals, according to them came from the same single-celled organism in the distant past; in their imagined "billions of years" ago.)

I watched Richard Dawkins via a video documentary posted on the Internet give a demonstration on how fish climbed out of water, turned into mammals, and then those mammals crawled back into the water and became whales. YES! he actually said that and actually made the presentation! I'm sorry, the notions evolutionists believe in are far more imaginary and unfounded than any statement made in the Holy Bible. Indeed! magical morphing cells that become bananas and the monkeys that eat them! that enable fish to become furry beasts and then return back into the oceans to become the largest creatures by far in all the world! (and yet these same people have the audacity to mock and laugh at Holy Bible believers)

Referring back to my observation: " mankind has learned that specific molecules make up DNA and that depending on how these chemicals are arranged results in the various genomes, but I have yet to read or see anyone demonstrate, how common elements and chemicals, sugar and proteins, result in the huge variety of life forms on earth. People talk about the hardware, essentially being hydrocarbons (molecules of hydrogen and carbon) and DNA being the software that manipulates these elements found in the earth and water to form living organisms. How? How do chemicals, "nucleotides and bases": * Adenine (A)* Cytosine (C)* Guanine (G)* Thymine (T) tell the elements found in water and the earth to make various complex living creatures?" How does merely the order of these substances form such massive varieties of life forms from simple to complex? How does ACGT in DNA not only take substance from the earth and water to make a single celled living organism, but to make our multi-trillion celled selves? Even using the search engines ever since the Internet came into existence, not one of the many papers I've read details answers to these questions, if anyone could, they would have been able to recreate a living cell from the raw materials and data, but to date no one has.

The reason in over a million pages of research in my lifetime these questions remain unanswered, in my opinion, can only be explained two ways 1) There is an outside Agency, Like God Almighty, that is actually in control of reality before our eyes and so forms and gives life to all manner of creatures or 2) There is yet more information, more details, in DNA or some other aspect of life that has yet to be discovered.

"All matter originates and exists only by virtue of a force which brings the particle of an atom to vibration and holds this most minute solar system of the atom together. We must assume behind this force the existence of a conscious and intelligent mind. This mind is the matrix of all matter."

- Max Planck -
https://www.goodreads.com/quotes/1328821-all-matter-originates-and-exists-only-by-virtue-of-a

The scriptures tell us the Intelligence that Created and Made all things and by which all things consist before our very eyes, is GOD, JESUS CHRIST. (a verifiable fact!)

Colossians 1:15-19

The Supremacy of the Son of God

15 The Son is the image of the invisible God, the
firstborn over all creation. 16 For in him all things were
created: things in heaven and on earth, visible and
invisible, whether thrones or powers or rulers or
authorities; all things have been created through him and

for him. [17] He is before all things, and in him all things hold together. [18] And he is the head of the body, the church; he is the beginning and the firstborn from among the dead, so that in everything he might have the supremacy. [19] For God was pleased to have all his fullness dwell in him,

6one God and Father of all, who is over all and through all and in all. https://biblehub.com/ephesians/4-6.htm

The Beginning

1In the beginning was the Word, and the Word was with God, and the Word was God.**2**He was with God in the beginning. **3**Through Him all things were made, and without Him nothing was made that has been made. **4**In Him was life, and that life was the light of men.... https://biblehub.com/john/1-3.htm

In short, the universe and life exists because our Eternal Creator exists. Intelligence likewise exists, because He is Intelligent. (Everything, especially everything Good, is evidence of the One True God, Jesus Christ: All Virtues, Truth, can be known because God is Eternally REAL and can be PERSONALLY KNOWN beyond all doubt.)

Either way, evolutionary claims of "goo to you by way of the zoo" are false and impossible; redundant DNA, specific to each kind of different creature in every cell of those respective kind of creatures, PREVENTS them from "evolving" into a different kind of creature. It's similar to backing up your operating system on your computer to make sure no virus or even desired software interferes with the smooth operation of it; only in our bodies such backup DNA redundant coding is TRILLIONS of times over! No evolution, NO WAY! no matter how much imaginary time passes! Instead, mankind should be very concerned about LOSS of DNA, and stop thinking on imaginary disproven evolutionary nonsense. We should be working on preserving quality DNA of genomes for each new generation or we sadly will see more and more birth defects and extinction of various kinds of creatures over time. (Repair any damaged DNA, genomes, by preserving good DNA, genomes, of each kind of creature; INSTEAD of altering them.) Hence, things like the seed vault are actually a very good thing.

What I mean by symbiotic relationships and the entire ecosystem of earth is interdependent is that if you remove "keystone species", the entire fabric of life on earth begins unraveling. If pollinators go extinct, so do the plants that rely on them, and whatever life-forms that rely on those plants. In other words, life on earth HAD TO

COME INTO EXISTENCE SIMULTANEOUSLY with ALL THAT'S NECESSARY for those living organisms to live or they couldn't! IT COULD NOT EVOLVE OVER DEEP TIME LIKE THEY ARE SAYING! IMPOSSIBLE! What we see throughout examining all the dead creatures of the past, is EXTINCTION; not evolution. In other words, life in all its complexity suddenly appeared on earth (what some refer to as the "Cambrian explosion"); but really, again, the fossil record shows marine organisms in all the sedimentary layers proving the flood; not deep time claims and when you understand that, it means all those creatures (of the purported geologic ages) existed together prior to that flood. Some might laugh when they read that for the first time. (Finding fossils all buried in sedimentary layers, especially when over 90% of those fossils are all marine organisms, is absolutely evidence of the historic flood! don't listen to anyone imagining otherwise!) When they claim those layers are hundreds of millions to billions of years apart, that's just RUBBISH!

How do you explain manmade artifacts found buried in coal seams that are supposed to be "millions to hundreds of millions of years old"?

https://www.google.com/search?client=opera&q=out+of+place+artifacts+found+in+coal+seams&sourceid=opera&ie=UTF-8&oe=UTF-8

I'm just pointing out that from soft tissue dinosaurs, to manmade artifacts buried in sedimentary layers supposedly hundreds of millions of years old, to inclusions of carbon in diamonds supposedly billions of years old; time and again evidence is ignored, flippantly dismissed by those "evolutionists" who desperately want to believe in their deep time claims and their magical morphing cells, instead of the most important and well verified facts of history in the Holy Bible.

There are many aspects not just evidence found on our planet, but of our own solar system that doesn't jive with deep time claims; just do some research on Spike Psarris astronomy and our solar system.

Look! I was force-fed evolutionary claims, same as you; so yes! I understand those claims! All I'm saying is that deliberate censorship of differing opinions is NOT beneficial to any of us! Creation Science SHOULD be in our schools, because creation scientists (both on the sides of young and old creation) bring up very valid points that are NOT being refuted by those who believe otherwise! Instead, they are just being unilaterally ignored and silenced; CONTRARY TO OUR MOST IMPORTANT RIGHTS, THE HIGHEST LAWS OF OUR NATION! If the American people still want to believe they came from pond scum and monkeys instead of God, then frankly, I'm so ashamed, I wish I could join a nation that's unashamed of Truth, Reality and Jesus Christ, the One True God! Like Americans used to be before antichrists wickedly removed the Holy Bible and prayer from our public schools! TRAITORS! enemies of AMERICA!

I repeat! virtually ALL of the nonsense, regarding the origin of our universe and life, being taught today wrongly in the name of science is not just FALSE, but IMPOSSIBLE!

https://www.facebook.com/photo/?fbid=6382847548460817&set=a.115635768515391 - PLEASE READ MY CITATIONS, AND THE CITATIONS WITHIN THOSE CITATIONS. I have plenty of valid reasons to support my conclusions!

I want to also reiterate that symbiotic relationships absolutely disprove evolutionary claims! you don't go from asexual mitosis to complex sexually reproducing multi-celled organisms by small changes over any amount of time! you HAVE to have FULLY functioning male and female sexual organs of all sexually reproducing creatures, OR THEY WOULDN'T EXIST! And if you don't have food present with any kind of life, it dies!

"Some pollination relationships are quite specific. A specialist relationship exists between the yucca moth and the yucca plant. Moth larvae only eat seeds of the yucca plant, and the plant depends on adult moths carrying pollen to facilitate sexual reproduction (Pellmyr *et al.* 1996). These specialist relationships can be negatively impacted if there is a decline of either species. A major determinant of Monarch butterfly population size is the availability of its host and nectar plant, milkweed, *Asclepias syriaca,* (Brower *et al.* 2011). Milkweed prevalence decreased between 1999 and 2009 in Iowa corn and soybean fields following the widespread adoption of herbicide resistant cultivars (Hartzler 2010)." - https://www.nature.com/scitable/knowledge/library/pests-and-pollinators-23564436/#:~:text=There%20are%20over%20200%2C000%20species,%2C%20solitary%20bees%2C%20and%20wasps. In other words, they've developed crops of corn and soybeans that can be sprayed with herbicides to kill weeds but not the food we eat. Who is testing these sprayed crops for how much of those herbicides and pesticides remain in/on them in the markets? With so many chronic and terminal ailments, digestive complications on the rise, it should be a standard of practice to ensure our food is safe for consumption.

The use of herbicides and pesticides, eliminates not just creatures identified as pests, but others that aren't. The more mankind focuses on genetically modified monocultures, routinely sprayed with poisons, the more life on earth as a whole is threatened. It's the intricate fabric of biodiversity, as designed by our Eternal Creator, Jesus Christ, that sustains life as a whole, including our own. Ignoring His Counsel in the Holy Bible only leads to suffering, death and destruction (extinction events). God tells us that because mankind is poisoning the earth, not just with their sinful ways but deliberately poisoning our own crops, they will have poison to drink. Jer 9:12-15 Disobeying God causes the curse to come upon our land, weeds and pests, (Deut 28) but rather than repent and learn to obey God, mankind is poisoning our food and as a result people are becoming sick because of herbicides and pesticides that are polluting/poisoning our food and water supplies. We must repent, those who grow crops must learn to speak in Faith as they Obey God, "In the Name of our Lord Jesus Christ, let this ground produce good harvest, good nutrition, and God deliver this ground from the curse from weeds and pestilences. Thank you, Lord for a bountiful harvest." You must follow what God tells us and speak over your children, your labors, your crops faithfully, providing what we all know they need, water, sun and fertilization, and also love and spiritual faith. All creation is more than just physical substance, it is supernatural in its essence and responds to the Words of God. It is why we all NEED GOD! People trying to do away with the mention of God and His Words in the Holy Bible are only headed for suffering, death and destruction.

I learned to speak over my plants as a child because my parents gave us our own garden patch and let us tend it ourselves. We knew that the ground had to be fertilized and the plants watered, and so we all had the same soil and water and sunlight. (everything physically needed to grow crops) But I gave my plants, something else. I hate to admit it, but as a child I took it as a competition and I wanted my garden patch to grow the best and produce the most, so I daily went out and spoke to the seeds, telling them to produce roots and get all the water and soil they needed to grow big and strong and produce a good harvest no matter what seed it was, beans to produce tons of beans, squash to produce abundantly and so on over my whole patch. I don't know if my parents realized it, but every year, my seeds sprouted first, grew the biggest and my patch produced the most. Because I loved those plants and spoke over them daily! I knew it was because of love and faith, when even in school we all had been given bean seeds from the same packet and all had the same soil, the same watering schedule, the same sunlit window, and yet my seed sprouted first, it grew the largest and produced more beans than anyone else in the class. It was because when other children went outside to play at recess, I stayed behind and loved up on my bean plant, speaking faithfully over it! **ALL CREATION**

RESPONDS TO LOVE AND FAITH, another MASSIVE EVIDENCE FOR GOD OUR ETERNAL CREATOR and THE TRUTH OF HIS WORDS IN THE HOLY BIBLE!

It should be noted that the presence of clean and unclean creatures, pests and pollinators, good and evil on earth, food and poisonous plants, etc. are all reasonably explained in the contents of the Holy Bible (not by the theory of evolution or anything espoused by God-denying atheists regarding their imaginations about how the universe and life came to be). In other words, observable reality has always only confirmed those contents of the Holy Bible as absolutely true.

Our Creator tells us that He Created and Made all living things and the text implies rather rapidly or none of those many various kinds of creatures would be able to live, feed and propagate from generation to generation. In other words, even today we see extinction instead of evolution happening before our eyes, which means life came into existence just as God tells us in the Holy Bible and has been dying off just as God tells us in the Holy Bible.

Matthew 24:35
Heaven and earth will pass away, but My words will never pass away.

https://www.esv.org/Psalm+102:26;Isaiah+51:6;2+Peter+3:10;Matthew+5:18;Hebrews+12:27/

When God plainly tells us that if we depart from Him, we will inevitably suffer, why would anyone in their right mind do so. So those who are advocating for the removal of the mention of God and the Holy Bible from our schools, are NOT in their right minds and are bringing suffering, death and destruction upon themselves and all the children they are lying to; in denial of God and His Commandments in the Holy Bible.

http://www.blastthetrumpet.org/PublicLetters/AAAUpdatedPublicAlertsMattersofLifeandDeath/Updates053016/Why%20We%20All%20Need%20Instructions%20From%20Our%20Creator%20to%20Live%20By.pdf - Yes, we NEED the Absolute Authority that comes from God, Jesus Christ, alone, or law and order fails, crimes and wars ensue and eventually, civilization ends.

http://www.blastthetrumpet.org/PublicLetters/AAAUpdatedPublicAlertsMattersofLifeandDeath/Updates053016/Too%20Many%20are%20Turning%20Science%20Into%20a%20Religion.pdf - people who idolize any scientist and their opinions over the contents of the Holy Bible are still practicing beliefs that are "religious" in essence, it's just in the form of a cult. When you believe ideas proposed by scientists that are simply false, then you have become a cultic believer in that scientist or scientists and are committing idolatry. **NEVER BELIEVE ANYONE OVER GOD ALMIGHTY, JESUS CHRIST, AND HIS WORDS IN THE HOLY BIBLE.**

http://www.blastthetrumpet.org/PublicLetters/AAAUpdatedPublicAlertsMattersofLifeandDeath/Updates053016/THIS%20IS%20ABSOLUTE%20SELF%20DESTRUCTIVE%20MADNESS.pdf - our laws do not protect criminals or cult members practicing crimes in any way. They are strictly for the lawful, the godly, moral citizens whose worldview is Biblically based.

http://www.blastthetrumpet.org/PublicLetters/AAAUpdatedPublicAlertsMattersofLifeandDeath/Updates053016/Religion%20in%20the%20Name%20of%20Science.pdf - censoring true science, because religious people support it, is how we get lies taught in our schools instead.

http://www.blastthetrumpet.org/PublicLetters/AAAUpdatedPublicAlertsMattersofLifeandDeath/Updates053016/SEPARATION%20OF%20CHURCH%20AND%20STATE%20IS%20IMPOSSIBLE.pdf - the notion that "separation of church and state" means you can't mention God in our schools or politics is a false and impossible one. (If individual fools what to delude themselves in denying our Eternal Creator, they cannot force everyone else to coddle such delusions! TRUTH is what needs to be taught in schools, even if that truth is deemed to be "religious"; not vain imaginations!) You either have the TRUTH, JESUS CHRIST, or you have purely imaginary beliefs in your head, everyone can be deemed to be cultic (idolizing someone over GOD) in their beliefs or religious, even atheists, as this note shows throughout.

http://www.blastthetrumpet.org/PublicLetters/AAAUpdatedPublicAlertsMattersofLifeandDeath/Updates053016/Original%20Intent%20of%20the%20Religious%20Clause%20of%20the%20First%20Amendment.pdf - as much as we might like to imagine that people must stop being themselves once they walk through school doors, or into politics, they don't and can't be expected to, without violating their own conscience. SO, GET OVER IT! America was founded by Christians and short of an all-out war, is a Christian nation! (In my frank opinion, if you don't like it - LEAVE!)

http://www.blastthetrumpet.org/PublicLetters/AAAUpdatedPublicAlertsMattersofLifeandDeath/Updates053016/HOW%20TO%20KNOW%20TRUTH%20IN%20A%20WORLD%20OF%20LIES.pdf - When people are led astray because they don't know the Living God personally, then they can be persuaded to believe almost anything, as evidenced by how many on earth believe in evolution today, even though it's false and impossible by everything mankind has already learned. Likewise, by how many people believe all kinds of things, even imaginary gods. People need to comprehend the REAL GOD showed Himself, Proved Himself, and told us all plainly how to KNOW Him PERSONALLY; so, no one has any excuse!

http://www.blastthetrumpet.org/PublicLetters/AAAUpdatedPublicAlertsMattersofLifeandDeath/Updates053016/JESUS%20the%20CHRIST%20is%20the%20Most%20Documented%20Person%20in%20the%20History%20of%20the%20World.pdf - Jesus Christ is the Most Documented Person in the entire history of the world; so, ANYONE claiming the Holy Bible is a work of fiction, myth or legend, is simply WRONG. I have to say it: such people are lacking so much accurate knowledge, that they should go back to grade school and demand a proper education! Our schools have failed these people by removing the Holy Bible from them and instead filling their heads with lies year after year!

http://www.blastthetrumpet.org/PublicLetters/AAAUpdatedPublicAlertsMattersofLifeandDeath/Updates053016/Denial%20of%20REALITY.pdf - when people graduate from our schools and universities and still don't know our Eternal Creator personally, it means public education has completely failed to teach those people the most important information on earth! FAILED COMPLETELY TO DO WHAT OUR SCHOOLS AND UNIVERSITIES WERE FOUNDED TO DO! And that is teach everyone not only *about* God and His Words in the Holy Bible, but to teach them HOW TO KNOW AND LEARN FROM HIM PERSONALLY!

http://www.blastthetrumpet.org/PublicLetters/AAAUpdatedPublicAlertsMattersofLifeandDeath/The%20Sufferings%20of%20Chris1.pdf - Observable reality, before our eyes today and throughout history, corresponds to the Words of God in the Holy Bible precisely. There's no getting around it, if we depart from Him and His Instructions therein, we suffer and die. Ever since we started lying to children in our schools, teaching them that they came from pond scum and monkeys, instead of God, our nation is increasingly suffering the exact Divine Consequences/Curses mentioned in Deuteronomy 28. (The whole world is, more and more, due to the fact that His Gospel has been broadcast worldwide, His Words in the Holy Bible published worldwide, and people everywhere are still refusing to obey Him and repent of their wicked ways.)

No sooner did Europe and America start turning from Him and His Words in the Holy Bible and antichrists have been invading! people just don't seem to comprehend; nations will erupt in violent conflict because you're bringing in people(enemies) with diametrically opposing beliefs! I don't want to see America or any once Christian nation in Europe turned into rubble and ruin, due to ignorance, treason and corruption within our borders.

Part of the reason antichrists have been succeeding in their evil agendas to corrupt our nations is due to individuals accepting bribes!

https://www.google.com/search?client=opera&q=universities+media+and+politicians+accepting+bribes+from+islamic+nations&sourceid=opera&ie=UTF-8&oe=UTF-8#ip=1 - when universities and television networks and governing authorities accept bribes to suddenly start smiling upon worldviews that attacked us, like 911, those are acts of treason taking place in our borders, and it's the same way antichrists like Rockefeller and other wealthy people ripped public education away from Christians in our nation. People, even scientists, accepting bribes to tell lies:

https://www.google.com/search?client=opera&q=scientists+bribed+to+tell+lies&sourceid=opera&ie=UTF-8&oe=UTF-8 - scientists tell lies! GOD DOESN'T! The Holy Bible has always been true and always will be!

And is why our schools are censoring facts that soundly disprove the claims of evolution.

https://www.google.com/search?client=opera&q=carbon+14+dating+tests+of+dinosaurs+censored&sourceid=opera&ie=UTF-8&oe=UTF-8#ip=1 - our schools are teaching lies to children, things that have been proven to be untrue.

https://www.google.com/search?client=opera&q=lies+in+textbooks&sourceid=opera&ie=UTF-8&oe=UTF-8#ip=1 - it's not just science that has been massively tampered with to the point of teaching children lies, but our history. Censoring the mention of God in schools REQUIRES PEOPLE TO LIE! Because God is mentioned throughout truthful history!

https://www.google.com/search?client=opera&q=lies+in+our+science+textbooks&sourceid=opera&ie=UTF-8&oe=UTF-8 - people need to understand that the wealthy have been scientifically studying ways to deceive and control them for generations. Patents on psy-ops, mass brainwashing, mind-control programming through television and so on are all very real.

So, when antichrists, wealthy people, with sinister personal agendas ripped education away from Holy Bible believers and American patriots and took control of the television networks, both tools have been used to deceive, control and corrupt us all. Their evil falls under the terms of "social engineering" or "population management" and it begins with lying to you about God Almighty, and His Words in the Holy Bible, in order to demoralize you with their own wicked and perverse desires.

(read my book on "Population Management")

My point is that you simply are not being taught things that are true in our schools today, especially regarding the origin of the universe and life on earth. Removing the Holy Bible is resulting in "dummying down" the population and brainwashing them so severely that they're resistant to facts and truthful information. We must put the Holy Bible and prayer back in our public schools and we need to protest all the perverse programming on television and in the movies. We need to return to being a godly nation, or America will more than likely be used as a Divine Warning to the rest of the world to never become so wicked!

If such wickedness continues and is allowed to spread, I am VERY concerned about the future of our nation. In nightmares, I saw some cities completely devastated, skyscrapers throughout turned to piles of rubble and ruin, death everywhere, toxic fumes, smoldering ashes; thick, dark smoke so prevalent you couldn't see the sky, the few, miserable souls still alive had gone insane; some staring blankly more dead than alive, others feeding on dead corpses, it was a kind of hell no Hollywood horror has ever depicted and I pray never happens, but that's the direction people are headed in if they don't repent and return to God, His Truthful Words and His Holy Commandments given us in the Holy Bible.

Again, the countless symbiotic relationships worldwide prove evolution is false and impossible. Fully formed creatures that are dependent on each other for their survival would NOT EXIST if they BOTH didn't come into existence together. So, evolutionists would have us believe not in small changes over time, but in massive simultaneous ones, such that we cannot see all the countless simultaneous symbiotic mutations in the fossil record at all or taking place anywhere in all the world today. The theory of evolution is not just false by everything observed, by everything we know to be true, by the facts of history and scientific laws that have lasted the longest, but upon close examination fails every imaginary step of the way through their "evolutionary tree".

Do you understand why the fossil record proves the claims of the theory of evolution is false?

1) marine fossils in sedimentary layers worldwide even on top of Mt. Everest validates the history of the global flood contained in the Holy Bible.

2) finding all manner of creatures buried in the same layers together, from single celled organisms all the way up to the most complex; again, validates the history of the flood not claims of evolution

3) finding LARGER versions of the same kind of creatures still living today proves devolution; the opposite of evolution!

So, the fossil record validates the historic flood and proves devolution and extinction events; not any evolutionary claims. watch it again, if you missed it the first time:
https://www.youtube.com/watch?v=_3CQtIWnx7c

Entropic Death - Creation Subject to Decay

19The creation waits in eager expectation for the revelation of the sons of God. **20**For the creation was subjected to futility, not by its own will, but because of the One who subjected it, in hope **21**that the **creation itself will be set free from its bondage to decay** and brought into the glorious freedom of the children of God. **22**We know that the whole creation has been groaning together in the pains of childbirth until the present time.... https://biblehub.com/romans/8-21.htm

Observable reality, the facts of history, all tell us the theory of evolution (along with the entire God-denying evolutionary paradigm of the universe from nothing that assembled itself, including the magical morphing cell that "evolved" over billions of years into all varieties of living creatures today), is false and impossible and one of those observations is that the universe, reality before our eyes is passing away in what some call is suffering an "entropic death".

So, the seventh reason that the theory of evolution is false and impossible is entropy. Yes, yes, I've heard those who argue, BUT LIFE... and about their completely fictional notion of the "cosmic inflation event", and my response to them is that if LIFE and all that comprises life flies in the face of any of these so-called scientific theories and laws, then that's an argument against those scientific theories and laws, and also an argument for our Eternal Creator, Intelligent Design. IN OTHER WORDS, EVERYTHING, AND I MEAN EVERTHING, MANKIND HAS EVER CREATED OR MADE REQUIRES INTELLIGENCE, AND IF YOU WATCH "HOW IT'S MADE" ON THE SCIENCE CHANNEL, YOU SEE THAT EVEN THE SIMPLEST THINGS WE CREATE AND MAKE REQUIRE MASSIVE AMOUNTS OF ENERGY, RESOURCES AND ENGINEERING, MASSIVE AMOUNTS! (for example just to make bottle caps, requires thousands of tons of aluminum, and to get aluminum requires tons of sediments in the earth that contain the elements that make aluminum and so massive amounts of excavation, and then smelting and refining, and then after making the aluminum it has to be compressed, shaped and molded, by all manner of labor and machinery until at the last millions of aluminum caps seal bottles for distribution all over the world.) So no, reality (the universe and everything therein) before our eyes did not create and make itself; no ifs, ands or buts. (the stars didn't assemble themselves, the chemicals, etc. that's ALL TOTALLY PREPOSTEROUS NONSENSE put forth by GOD-DENYING FOOLS who just don't want to take God's Word for how He created and made everything in the Holy Bible!)

God made everything with its own seed for its own kind. In the beginning of creation, there were huge varieties of animals, mankind lived longer. It wasn't until later generations did we live shorter. https://www.youtube.com/watch?v=_3CQtIWnx7c&t=581s As Trey Smith aptly points outs. THAT IS DUE TO THIS CREATION BEING SUBJECT TO DECAY! None of it points to evolution! It is all subject to decay!

Let's consider one of the things mankind has observed about that reality; regarding "entropy". Particle physics, chemistry and all the sciences, owe their existence to God Almighty, Jesus Christ, and His Words in the Holy Bible. Scholarship began with those who studied what our Eternal Creator has been teaching mankind since the Beginning and those Divine Instructions are found in the contents of the Holy Bible. It was His Commandments that tell us to STUDY, to OBSERVE, to treasure Knowledge and Wisdom, that founded monasteries of early learning and scholarship. From those monasteries arose the first universities, and while generations of scribes wrote scrolls of knowledge for the coming generations, eventually those Biblical scholars realized that we needed a better way to reach the people of the world and so the movable type printing press was invented.

The Gutenberg Printing Press was invented to print the Gutenberg Holy Bible, the first of the kind of modern books we have today. As a result, books came into western civilization, followed by the age of reason and enlightenment, the founders of the modern sciences.

But that wouldn't have been possible, if not for the Church of Jesus Christ, the brave crusaders that fought off raping, robbing, enslaving, murdering hordes of muslims that had invaded Europe from the hell hole they crawled out of in Arabia. So, America and our most treasured freedoms and rights, owes its existence to Jesus Christ, and His Divine Commandments found in the Holy Bible. Advanced sciences and technologies have not come into this world through any of the other worldviews throughout history. Every claim by muslims that their cult has brought anything good into this world, is nothing but a lie. islam is one of the most barbaric, evil worldviews on earth and their claims of scientific knowledge came from the rapists, robbers, slavers and murderers that raided civilization (from the peaceful Biblical Commandments of God) and in the process stole not just materials such as silver and gold, but knowledge, taking what scrolls they found in ancient libraries, before burning and destroying such edifices of antiquity. The dark ages are often portrayed as a criticism only of those who held power over the people because they were keeping knowledge to themselves rather than educating them, but much of the dark ages was actually due to the darkness of islam as it spread its evil over the earth all the way into Europe before being repelled by the Crusaders. So, if not for the Christian Crusaders, Europe would have fallen to the darkness of islam in those dark ages and the western world would have been under the same hellish oppression, ignorance and wickedness islamic nations have endured all this time. Some islamic nations are so evil they ban the Word of God, the Holy Bible. The only reason islam prospers at all, is the rest of the world has been buying crude from them, making those nations rich, otherwise, islam would have died off simply due to droughts and famines and due to the fact, it's dogmas and doctrines are ones of death, enslavement and oppression; instead of life, liberty and freedom. Letting the ignorance and wickedness of the criminal organization of islam into western Christianized nations, is an open act of treason, spitting on the blood of our ancestors who fought to the death to keep that evil out; if allowed to continue will only mean suffering, death and destruction for sadly, so many souls.

(Read the book, "Save the World from islam" islam is a criminal organization and the whole world needs to be saved from it including muslims.) This recent betrayal and treasonous from our leaders to islam (namely their love of OPEC money), is causing atrocities to be committed against innocent people, Christians murdered and driven from their homes into refugee camps by the millions, while the rest of the world is paying little attention to their plight.

So, as I was saying, the sciences owe their existence to God, Jesus Christ, the contents of the Holy Bible and those who learned from God to study, learn, and treasure Wisdom and Knowledge and that God commanded all children to be taught His Precepts found in those Sacred Texts. (Public Education in western civilization began in the Church of Jesus Christ Triumphant; in other words, with the people who studied and read the Commandments of God in the Holy Bible.) And so, with the advent of modern books, Divine Knowledge spread, which is why virtually all the founders of the sciences were Biblical Monotheists, (believers in Jesus Christ as the One True God that He is) and in their notes credited Him and their observations about His Creation.

http://www.blastthetrumpet.org/PublicLetters/AAAUpdatedPublicAlertsMattersofLifeandDeath/Updates053016/Fathers%20of%20the%20Sciences%20and%20Great%20Scientific%20Minds.pdf - (cite quotes from the founders of the sciences and Nobel Prize winning scientists)

https://mathshistory.st-andrews.ac.uk/Biographies/Clausius/ - " **Biography**

Rudolf Clausius's father, Rev C E G Clausius, was a Councillor of the Royal Government School Board. He founded a small private school, becoming its principal, and as a minister of the church he also served as its pastor. "

Morris, Henry. *Men of Science Men of God : Great Scientists Who Believed the Bible, revised edition*. Master Books. 1988 The vast majority of the founders of the fields of the modern sciences and Nobel Prize winning scientists have all been believers in Jesus Christ and His Words in the Holy Bible. So true science is not in opposition to the Words of God in the Holy Bible at all. (Any scientific opinions that are contrary to those Divinely Inspired contents are false and should be discarded.)

https://www.icr.org/article/genesis-question-entropy - this present creation is suffering "an entropic heat death", that is over time is losing heat and approaching a state of equilibrium as energy and matter continues to dissipate. So, since this universe had a Beginning, in which the oldest form of energy observed to date is called "cosmic background radiation" and is passing away, then everything we have observed and know about creation confirms the Biblical Account. Any BELIEFS even in the name of science (from scientists) that are CONTRARY to the Words of God in the Holy Bible are false (like the theory of evolution and the entire evolutionary paradigm of the universe from nothing, of the mass and energy that assembled itself but is dissipating and cooling off (entropy) before our eyes.) So, entropy tells us that their evolutionary theory of the universe creating and assembling itself, including life, including their magical morphing cell that "evolved" over time, flies staunchly in the face the 2nd Law of Thermodynamics and is therefore FALSE and IMPOSSIBLE.

Over time, all that has been happening on earth is death, decay, erosion, entropy. Anything that we create, make or build (any order arising from the ashes of death, decay, erosion and entropy) is directed, intelligent energy at work to assemble, create, make and build anything; which points irrefutably to Intelligent Design of Creation by Our Eternal Creator.

There is simply no way around the fact that creation exists by the Power and Intelligence of our Eternal Creator, Jesus Christ. This fallen creation, INCLUDING LIFE, has been subjected to DECAY! We are not evolving at all. As time goes on more and more genetic information is lost. NOT EVOLUTION BY A LONG SHOT!

It's from the observation of the universe expanding and cooling off, "entropic heat death" and the many extinctions of living creatures on earth, that we know that what God tells us in the Holy Bible is true.

35 Heaven and earth will pass away, but My words will never pass away. https://biblehub.com/matthew/24-35.htm (read the cross references as well)

So, reality before our eyes, every form of scientific knowledge that is true science, exposes the stories being taught today in the name of science (regarding the origin of the universe and life on earth) are all false. The ONLY WAY creation (universe and life) exists is by the Power, Will and Intelligence of our Eternal Creator. (And yes, I've heard all the critics who say things like then why are we suffering? why are we dying, why is this universe passing away? the answers to their questions are found in the Words of God so openly published worldwide in the Holy Bible; that they mock and laugh at without reading, without studying; without learning. In other words, all such persons are arguing not from a state of knowledge, but from a lack of it.)

Everything mankind learns by studying creation, tells us that the God-denying story of how the universe and life came to be is false and impossible, and also tells us that what God has told us in the Holy Bible is absolutely true from cover to cover.

For example, particle physics. I tend to think Chuck Missler was someone in recent history that was showing a proper connection between the Truth contained in the Holy Bible and what we observe about reality before our very eyes, I really wish more people listened to his presentations.

https://www.youtube.com/watch?v=X_AeNHDv208 and
https://www.youtube.com/watch?v=o8nHYa0iSrg and
https://www.youtube.com/watch?v=HqkJ_CUWz2I

In other words, the smallest particles that are found in everything in our universe have some incredible properties. (Supernatural as opposed to natural.) We essentially exist and are part of Supernatural reality as opposed to natural. In other words, the universe isn't just Supernatural in Origin but in Essence. The scriptures put it this way:

Paul Before the Areopagus

...**27**God intended that they would seek Him and perhaps
reach out for Him and find Him, though He is not far from
each one of us. **28**'For in Him we live and move and have
our being.' As some of your own poets have said, ' We
are His offspring.' **29**Therefore, being offspring of God, we
should not think that the Divine Being is like gold or silver
or stone, an image formed by man's skill and
imagination.... https://biblehub.com/acts/17-28.htm

AND

Colossians 1:15-20

The Preeminence of Christ

15 He is the image of the invisible God, the firstborn of all creation. **16** For by[a] him all things were created, in heaven and on earth, visible and invisible, whether thrones or dominions or rulers or authorities—all things were created through him and for him. **17** And he is before all things, and in him all things hold together. **18** And he is the head of the body, the church. He is the beginning, the firstborn from the dead, that in everything he might be preeminent. **19** For in him all the fullness of God was pleased to dwell, **20** and through him to reconcile to himself all things, whether on earth or in heaven, making peace by the blood of his cross.

AND

3By faith we understand that the universe was formed at God's command, so that what is seen was not made out of what was visible. https://biblehub.com/hebrews/11-3.htm

AND

Genesis 1:1

In the beginning God created the heavens and the earth.

Psalm 33:6

By the word of the LORD the heavens were made, and all the stars by the breath of His mouth.

Psalm 33:9

For He spoke, and it came to be; He commanded, and it stood firm.

John 1:3

Through Him all things were made, and without Him nothing was made that has been made.

…**19**For what may be known about God is plain to them, because God has made it plain to them. **20**For since the creation of the world God's invisible qualities, His eternal power and divine nature, have been clearly seen, being understood from His workmanship, so that men are without excuse. **21**For although they knew God, they neither glorified Him as God nor gave thanks to Him, but they became futile in their thinking and darkened in their foolish hearts.... https://biblehub.com/romans/1-20.htm

Essentially Logic, Mathematics and the Sciences ALL tell us that the theory of evolution is false and impossible! And instead, overwhelmingly confirms the contents of the Holy Bible. (The universe/reality is Supernatural both in origin and essence.) - Some of those who have stated evolution is false: Michael Denton, Philip Johnson, Michael Behe.

"The theory of evolution is not supported by thinking people." - Chuck Missler

https://www.youtube.com/watch?v=aeEuwrDVQ98 and (again) https://www.youtube.com/watch?v=HqkJ_CUWz2I

Missler makes some statements in his presentations that requires pausing to think about. Like his statements about quanta, sizes and distances between the nucleus of an atom and the electron cloud surrounding it and other properties of particles that comprise reality before our eyes. Quanta being so small that essentially locality is lost (the particle virtually is no longer observable or measurable because it becomes so small and one of the reasons Einstein's Theory of Relativity is mistaken.)

https://www.youtube.com/watch?v=PZ3hESj__M8 - when I refer people to other sources of information in this world, it doesn't mean I agree with EVERYTHING in the citation. Like when I tell people they should listen to Chuck Missler and his perspective of the Holy Bible in 24 hours (24 - 1-hour presentations), it just means that out of all the books I've read, and people I've listened to, Chuck Missler is really thinking about HOW REALITY (scientific observations) are inextricably tied to the Words of God, our Eternal Creator, in the Holy Bible. That it really is God, Jesus Christ, and His Words that are responsible for reality before our very eyes.

Another person I've encountered who is thinking about reality, the facts of history and the Truth that God plainly tells us all in the Holy Bible is Trey Smith. So, I recommend watching Trey's presentations as well. Again, NOT THAT I AM AGREEING WITH ANYONE ELSE BUT GOD ALMIGHTY ON EVERYTHING, just that I have found some people are much closer to grasping the "big picture" and say things that we all should really give serious thought to.

https://www.youtube.com/watch?v=_3CQtIWnx7c - Virtually all his presentations and books, have information that people who have been deceived by the current nonsense in our schools and universities really need to examine much more thoroughly than they ever have before in their lives.

I've seen people laughing at Ken Ham and ICR and Kent Hovind even though when they do that, they are making themselves look like brainwashed morons who've lost the ability to examine information provided to them! STOP DOING THAT! If you can't refute information given you, then you need re-examine your current beliefs and a refutation isn't parroting a bunch of lying nonsense you've heard from anyone else. A sound refutation against any information given you must be truthful. (It's why I cite the Words of God Almighty as that absolute Foundation of Truthful Information given mankind! THERE IS NO GREATER AUTHORITY THAN GOD, JESUS CHRIST! so if you cite an opinion of anyone else, even some scientist as your authority, it means you are guilty of idolatry and are calling God Almighty a liar! it's very evil to believe ANYONE over GOD, Jesus Christ, and His Words of Truth in the Holy Bible!)

When Kent Hovind and other Biblical Creationists tell you point blank that you are being lied to in our schools about a great many things, it's because you are! And when you laugh in return, it means you have been factually brainwashed to believe those lies.

https://www.google.com/search?client=opera&q=hovind+grand+canyon&sourceid=opera&ie=UTF-8&oe=UTF-8 - when a story like the Colorado River carved the Grand

Canyon over millions of years is disproven by evidence and facts of history, then it needs to be discarded!

https://www.google.com/search?client=opera&q=hovind+lies+in+our+text+books&sourceid=opera&ie=UTF-8&oe=UTF-8 - when there are still proven lies and hoaxes being taught children, those are crimes against humanity, BRAINWASHING/DEMORALIZATION! so severe that it has people laughing at the well-researched and verified contents of the Holy Bible and calling it fiction, myths and legends is making those people look like absolute idiots! It is resulting in all manner of evil!

https://archive.org/details/DarwinsDeadlyLegacy - lies being taught children in our schools is absolutely a form of abuse and it results in all manner of evil in this world.

Everyone needs to realize that we must be concerned about WHAT IS TRUTH over whether or not TRUTH is "religious". In other words, removing the Holy Bible from our schools, immediately removes TRUTH from our schools and inevitably results in things that are not true being taught children. **The facts of all recorded history are inseparable from God who is mentioned throughout! BECAUSE GOD IS REAL! So NEVER and I mean NEVER let God-denying antichrists take over public**

education! it means that lies will inevitably be taught innocent children, brainwashing them so severely that they are acting like absolute idiots today! Insisting they're direct descendants of monkeys and pond scum over billions of years and laughing like hyenas at the very mention of God. Brainwashed Boneheads! and I'm not saying that to insult anyone, just like God is saying those who deny Him are fools is an insult, it's just true by the definitions of the words. In other words, **put the Holy Bible back in our schools and use it thoroughly daily, as the primary textbook for accurate information in this world; so people will stop presenting themselves as brainwashed boneheads; fools on their way to the flames of damnation** and so that much good will result in this world instead of so much evil.

On the one hand, scientists tell us that gravity pertains to mass/matter and on the other tells us that photons are "massless particles". If light and its photon particles are massless, how can gravity have any effect on it? how can light be bent, reflected, refracted, diffused, slowed and even stopped by other forms of mass and energy? So, photons might be extremely small, but still must have some mass to them. The errors in relativity are not just the wrong assumptions about light and the various speeds it travels, definitions of mass and energy, but about time.

Universal Time is the measurement of the rate of the expansion of matter and energy from the Beginning of Creation. It is a DESIGNATED increment, the smallest of which would be quanta so small that for all practical purposes wouldn't be measurable. In other words, the designated increment needs to be large enough to be measurable. But understand Time is DESIGNATED, purely an intellectual construct, and so cannot be affected by mass or energy. (Relativity is mistaken, along with the many notions that have come from it.) I talk more about mapping space and time (objects therein) in some of my other materials and videos. (Passing on what God taught me about the Fourth Dimension, Space and Time, Reality, Creation.)

(If time permits, I'll write a book on it "Mapping Space and Time, the Fourth Dimension, the House of God") Most all of the theories about prehistory is all just as wrong as evolution! IT'S ALL FAULTY IMAGINATIONS! It has NOTHING to do with reality!

Particles that all respond to Light. In other words, without Light we can't observe them, without Light, (the electro-magnetic spectrum) the particles don't possess the properties necessary to become reality before our very eyes. It's the reason some are saying we are living in a holographic universe. Which is another way of saying that our Intelligent Creator is not just the Originator of

Creation, but is the Presenter of it; controlling reality before our very eyes moment by moment.

If you want to consider what I'm saying at this point to be a Theory then call it the Theory of a Supernatural Universe - The Universe is not just Supernatural in Origin but in Essence.

You have to really think long and hard about particles that comprise all matter and relate to all forms of observable energy. The four states of matter with the various properties, and how everything we see is comprised of these particles that merely by the number of them in any given space, change their appearance from plasma (forms of Light) to gases, liquids and solids, all manner of elements. How these charged particles adhere and combine electromagnetically, chemically into molecules that are the building blocks of the entire Creation before us all, including Life. Just to observe these particles and to begin to try to understand them, took thousands of years of advancements in technology among mankind. And when we finally succeeded in observing them, those particles are changed JUST BY LOOKING AT THEM. (I chuckle a little inside, because it reminds me of those videos where people use a laser pointer to play with their pet kittens.)

https://www.google.com/search?client=opera&q=the+observer+effect+explained&sourceid=opera&ie=UTF-8&oe=UTF-8 - the observer effect is something we can easily know just by the presence of each other, especially when watching any television program or movie. In other words, people do things all the time merely because they know someone else is looking at or watching them. Interactions like that are not just observed macroscopically, but microscopically. Which means it takes INTELLIGENCE to observe an INTELLIGENT Creation, that is most definitely Engineered to be Observed and Understood!

So, the properties of particle physics are no friend whatsoever to the evolutionary paradigm and instead coincides exactly with what our Eternal Creator plainly tells us in the Holy Bible.

https://www.khouse.org/articles/2012/1086/

One of the very first things God tells us in the Holy Bible regarding Himself and Creation is, "Let there be Light." It's not just the observation of cosmic background radiation, but that Light is found throughout creation. (not all forms of Light are visible; the vast majority of the electro-magnetic spectrum (Light) isn't visible.) But without Light the universe as it exists today wouldn't exist at all. No wonder God tells us that He is Light. (one of His Many Identifying Statements about Himself)

Psalm 139:12

even the darkness is not dark to You, but the night shines like the day, for darkness is as light to You. (listen again to what Missler says about plasma throughout our universe - https://www.youtube.com/results?search_query=Chuck+Missler+our+holographic+universe)

Walking in the Light

4We write these things so that our joy may be complete. **5**And this is the message we have heard from Him and announce to you: God is light, and in Him there is no darkness at all. **6**If we say we have fellowship with Him yet walk in the darkness, we lie and do not practice the truth.... https://biblehub.com/1_john/1-5.htm

To be sure God tells us His Words are Spirit and Truth. Sometimes, I see people misunderstanding Him and His Words because they are leaning in one direction or the other. Certainly, God defines His Own Terms in the Holy Bible (gives His Own Divine Definitions of His Words) and as such you can never go wrong if you stick to those given Divine Definitions. When I say one direction or the other, I mean people try to comprehend the contents of the Holy Bible, without those Divine Definitions and bring their own definitions and connotations of words. That results in misunderstanding those contents. To be sure the Words of God govern both Spiritual Truth and Reality, the Visible Universe, but if you bring a head full of lies, wrong definitions of words, wrong connotations about reality to the Holy Bible then you won't understand those contents. People refer to people as wearing a subjective pair of glasses that prevents them from objectively, truthfully understanding anything correctly. I simply say that far too many are still lacking correct information (ignorant), and are still deceived (believing lies) and as a result have wrong imaginations about reality and the Holy Bible (self-deluded.) So, these people might imagine they are in their right minds, but they aren't.

When someone laughs at the most thoroughly researched and verified book in all the world, telling us the absolute truth about the past, present and future, the most important accurate knowledge in all the world, the provably and verifiably Divinely Inspired contents of the Holy Bible, they epitomize what I just stated about being ignorant, deceived and self-deluded; so much so that to the sane and properly educated, they appear not to be at all.

I see ignorant, yet arrogant, disdain for Biblical Creationists; despite the fact that the Holy Bible is thoroughly verified by everything mankind knows to be true; from people who have no rationale whatsoever for reality before their very eyes! atheists laugh at and mock Holy Bible believers daily, not realizing that it makes them look insane, ignorant, deceived and self-deluded! While they imagine it's the theists who are lacking knowledge, it's them! a head full of nothing but lies! and it comes out in everything they say or write: lies, lies and more lies (like "The Universe from Nothing"). Most all of them believe the entire pile of crap, the evolutionary paradigm, that was ingrained into them throughout all their years of coerced indoctrination by God-denying fools today. I say fools, not so willingly on my part, but because God Himself says so and from my point of view the reasons He says so, is that He wants us all to stop being foolish, and wants those who are not being foolish presently to not choose the path of following (learning from) any God-denying fools, who practically throw temper tantrums if you even mention the Holy Bible and the One True God, Jesus Christ. (Agnostics are at least honest enough to admit their own ignorance, in that if they don't know God yet, it doesn't mean He doesn't exist, similar to the fact most people have never met famous actors and actresses but are not so foolish to claim those real people don't exist.) **If someone doesn't know God, Jesus Christ, they have no authority to teach anyone about anything.**

When I encounter people claiming the Holy Bible is a work of fiction by men, the thoughts that go through my head instantly (ignorant, deceived, self-deluded) and if I tell them the many reasons why the Holy Bible is provably and verifiably Divinely Inspired Truth, and they STILL insist on their madness of willful denial, then I know those are people who are WILLFULLY ignorant, stupid on purpose, and so much so that they are the definition of a self-destructive stupid fool on their way to the flames of damnation; acting like they have not one concern for their own souls. I use hard words not to insult anyone, but because some people have let the devil turn them into hard heads and hard hearts and are not using the mind God has given them in any productive, proper way. Such God-denying antichrists are what's wrong in this world, cancerous parasites that accept the Goodness of God daily (their existence, every breath, food, water and all the necessities of life), while perpetuating evil on earth.

We are commanded by God, Jesus Christ, to proclaim the Truth, His Words in the Holy Bible to everyone. Anyone WILLFULLY RESISTING HEARING OR READING HIS WORDS, is absolutely showing at the very least they are completely enslaved by the very real devil, by his unholy spirit of err/deception/antichrist and at worst is a child of the devil in the flesh, headed for the Wrath of God and the Lake of Fire. If anyone WILLFULLY refuses to listen to or read the Words of God in the Holy

Bible, shake the dust off against them! Don't waste time and energy on self-destructive, stupid persons, who are headed for the flames of damnation, instead move on to seeking lost souls, who will rejoice at being Found!

https://www.facebook.com/photo.php?fbid=6169704459775128&set=pb.100002069048072.-2207520000&type=3

When I say self-destructive, evil ends itself, as evidenced by violent conflict throughout history. Hating each other, resulting in abuse, torture, murder, wars, etc. means that if everyone chose to be evil, people would inevitably kill each other down to the very last. So, God, Jesus Christ, tells us the Truth that to live, we must learn to Love Him and each other instead. He also tells us that He is not going to allow the devil and evil people to endlessly attack those who see the Wisdom of learning to Love Him and each other. So, such wicked persons will be ultimately confined in the lake of fire. People should stop acting so hypocritically for faulting God with what lawful citizens do to repeat violent offenders by imprisoning them for life or executing such persons who continue to be so thoroughly evil even behind bars. In other words, if anyone REFUSES to Obey God's Commandment to REPENT of our wicked thoughts, words, ways and deeds, He will imprison them; some say destroy in the lake of fire.

I repeat, anyone who refuses to acknowledge even the POSSIBILITY of our Eternal Creator is demonstrating a state of mind that is so willfully ignorant, deceived and self-deluded; that it's synonymous with insanity; denial of the most important facts of history and reality before their very eyes. It's the reason atheism used to be considered a form of mental illness and frankly, still should be.

http://www.blastthetrumpet.org/PublicLetters/AAAUpdatedPublicAlertsMattersofLifeandDeath/Updates053016/Acknowledging%20the%20Eternal%20Creator%20Takes%20No%20Faith%20It%20is%20Scientific%20Fact.pdf - Acknowledging the existence of our Eternal Creator takes no faith, it only takes accurate knowledge and a reasonable mind. What takes Faith, is approaching Him; believing Him and His Words above all.

Some imagine that the devil and all unrepentant persons are completely destroyed in the lake of fire. I personally am not so certain that wicked beings are completely, immediately destroyed, just because of what God says about them in the Holy Bible (verses like "**10**And the devil who had deceived them was thrown into the lake of fire and sulfur, into which the beast and the false prophet had already been

thrown. There they will be
tormented day and night forever and ever. " Rev 20:10).

Some people just don't get it! God describes Himself like a Strong Tower that the Wise take refuge in. If you are outside of Heaven and instead are confined in the lake of fire (Rev 20:15; 21:8), you're outside of everything Good and Perfect, Creation Perfected and, in a place, where only evil exists, wicked spirits, unclean, defiled souls, their own evil and shame, their own realization that they could have gone to Heaven but refused to repent of their wicked ways, automatically causes them torment. It should be noted by Rev 3:5 that we ALL were given life from our Eternal Creator and it's His desire that we learn to exist with Him and each other in Love, Peace, Unity, Harmony and all Virtues together (called the Fruits of His Holy Spirit - Galatians 5:22-23 **22**But the fruit of the Spirit is love, joy, peace, patience, kindness, goodness, faithfulness, **23** gentleness, self-control; against such things there is no law.) But if you choose to REFUSE to live righteously and instead choose to live wickedly, you will be confined in the lake of fire; prevented from attacking any of His Loving, Peaceful, Wise Sons and Daughters ever again. And no God, Jesus Christ, isn't evil for arresting evildoers and telling us all that there are Divine Rewards and Consequences based on our choices and actions.

In this world, provided you choose to live lawfully, you will encounter other people choosing to live lawful, orderly lives and get along peacefully with each other. But should you encounter someone who is behaving unlawfully, disorderly, chaotically, irrationally, unreasonably, you experience evil in that encounter. Lawful persons arrest these criminals; where in prison, if the criminal doesn't repent but goes on committing crimes, then criminals suffer from the evil of other criminals present; so anyone who refuses to obey God, Jesus Christ, will not enter His Kingdom, will not be allowed into Paradise and instead will be confined in the lake of fire, where they will suffer just because they are in a place devoid of the Goodness of God, because they deliberately rejected Him and His Righteous Commandments to Love Him and each other and insisted on being evil instead.

In other words, order does not arise from disorder, chaos; it's why we arrest disorderly people, and why our immune systems kick into overdrive against any cells not functioning properly within our bodies. (chaos leads only to death and destruction) Entropy disproves the entire evolutionary paradigm of how the universe came from nothing and assembled itself, including the theory of evolution, concocted by people who are denying our Eternal Creator. Everyone is a Divine Creation with Divine Purpose, not a random "evolved" organism from monkeys in the distant past or pond scum prior to that. The universe did not assemble itself and become ordered all on its own from an explosion or an expansion, "big bang", of anything. From the tiniest particles to the most complex structures, it's all Intelligently Created and Made and responds to intelligent observations and deliberate constructs accordingly. (skyscrapers do not assemble themselves, jet engines do not spontaneously appear, and neither do our much more complex, multi-trillion celled bodies form themselves). **EVERYTHING created and made is the result of ENGINEERING, Intelligent Design. Creation is overwhelming evidence of our Eternal Creator.**

He tells us in the Holy Bible that we are much more than just bodies of flesh and bone, physical forms, but that He BREATHES LIFE into each of us, the word for that Divine Breath in ancient Hebrew is also the Word for spirit. And so, when I died as an unbeliever, my heart had stopped beating, the monitor coded in the hospital showing no heart beat or brain function (no electrical impulses activating vital organs), and my spirit had come out of my then dead body. In other words, people don't cease to exist after "death", just as God plainly tells us in the Holy Bible. NDEs have verified that fact redundantly now. What's labeled "Near Death Experiences" are really what used to be more aptly be called "Beyond (death) and Back" experiences. And no, these experiences are not explained by claims of hallucinations or latent brain function at all. Like myself, people testify of seeing and hearing things

AFTER dying that REALLY happened, witnessing conversations people had after they had died in completely different areas from where their dead bodies were, seeing things even on roof tops that were only verified after people who were resuscitated told others that object was there; even though access to that roof was behind locked doors. In other words, hallucinations and latent brain function after death does not explain the data collected by researchers on this topic, at least not all the many accounts to date. In addition, it should be noted that of those many accounts, people from all worldviews, including atheism, come back telling us they saw Jesus Christ, the One True God, and these people have been spending the rest of their lives telling everyone that the Holy Bible is telling us all the truth - persons like myself.

https://www.facebook.com/photo.php?fbid=6769993359746232&set=pb.100002069048072.-2207520000&type=3

I know it's hard to face that the reason people don't know the Truth, don't know God, Jesus Christ, is due to the fact they're believing lies, but that is indeed the Truth. When I died as a young man and ended up in outer darkness (it's a void of nothing beyond the furthest galaxies), God confronted me Himself. Conveying from His Existence to mine, "Alright wiseguy! If I don't exist, can YOU bring the sun, moon and stars into existence? can

YOU bring ANYTHING (even an atom or a quark or smallest particle(s), by implication) into existence from yourself (by my own will and power) from this void of nothing?

In that confrontation the very real Eternal Creator taught me two things virtually instantaneously: 1) There most definitely is a very real God who has THAT KIND OF POWER none of the rest of us have! The Power to Create Reality just by speaking, just by His Own Will, Intelligence and Power! and 2) I am not Him! (I don't have the Power to just speak and create the universe or even the tiniest particle in it from nothing but my own spiritual existence) When we die, we leave our physical bodies that are made from the dust and water of the earth and our spirits still have our thoughts, emotions, memories, while our physical bodies return to the earth from which God formed them. As one of the many deceived persons from years of public education I died as a young man; forgetting that I knew God even as a very young child and was instead accepting all those many lies currently being taught in our schools today. And how that it was somehow foolish to believe God Almighty even existed. (It's a tactic of brainwashing, which is why these people who have been subjected to it, and harmed by it, sneer at Holy Bible believers, even though they have no rationale whatsoever for reality before their own eyes. They have factually been brainwashed/demoralized.) You see the same kind of behavior from muslims, who say a person is "insulting" them just by telling them the truth: that islam has taught them so many lies, they can't seem to tell good from evil, God from the devil, fact from fiction or truth from lies anymore. (brainwashed)

(read my book Save the World from islam)

I repeat, that since all particles respond to intelligent control, form reality before our eyes, such things as crystals, elements, molecules, stars, planets, etc. etc. then even the tiniest things we can observe tell us they were Intelligently Designed! ANYTHING that explodes: disintegrates, goes the way of entropy, so the "big bang" theory FAILS IMMEDIATELY! NO! REJECT NONSENSE! Since the universe obviously began in a superheated state (if not we wouldn't see the blazing hot stars all over the heavens), by evidence of cosmic background radiation and plasma throughout, we know the current evolutionary paradigm is FALSE! No ifs, ands or buts! It's NONSENSE! You simply cannot get around the fact that creation exists by the Power of the Very REAL ETERNAL CREATOR who plainly tells us how He Created and Made everything in the contents of the Holy Bible. I don't fault anyone for asking Him HOW, desiring to the know the details, but I do fault anyone for attempting to deny Him and stepping into the realm of imaginary nonsense and calling those irrational imaginations "science".

There are books outside of the compiled contents of the Holy Bible, some call pseudepigraphal or apocryphal works, essentially these books are controversial as being Divinely Inspired. I've examined numbers of them and find that in my own opinion some are and some aren't. In other words, God tells us that He and His Words are TRUTH, any additional information in all creation STANDS OR FALLS upon the FOUNDATION LAID BY GOD HIMSELF, the contents of the Holy Bible! So, there are other books that go into additional details about those contents, without detracting from those contents, without denying those contents, or in any way opposing those contents. I personally don't encourage anyone to read such books UNTIL THEY KNOW GOD PERSONALLY and have been studying the contents of the Holy Bible from Genesis to Revelation for a number of years! ESTABLISH YOURSELF IN TRUTH SO THAT YOU CAN EASILY RECOGNIZE LIES!

Anyway, I'm not saying that God showed me exactly how He created this present universe, but He showed me how He could create a New Heavens and earth and it made a lot more sense than anything I've heard or read from anyone else.

(hopefully coming in the future, my book on mapping space and time – I will explain, in that book, how the sacred geometry defines the coordinates in the space-time continuum from the Beginning and how that everything that exists within space-time relates to those coordinates that space-time originates from the Beginning of the Creation of the Heavens and earth and that there is actually a center/origin of space time similar to the center of a sphere. BECAUSE that's HOW God showed me, He could create a New Heavens and earth, a MASSIVE BURNING SPHERE of ALL MATTER AND ENERGY COMBINED AND THEN SPUN BY HIM such that the burning mass and energy was flung outward and became the galaxies and stars, stretched by Him and His Hands! SHOWS FORTH HIS HANDIWORK! STRETCHES THE HEAVENS! God can make His HANDS ANY SIZE HE WANTS! HE CAN HANDLE BURNING MASS AND ENERGY WITHOUT ANY HARM TO HIM BECAUSE HE IS GOD!)

So, what I saw explains the universe better than any theory I've heard or read from anyone. I'm not insistent that this universe was made in the way God showed me He could make a new one, but I am insistent that the Holy Bible is telling us the truth and that anyone who expresses an opinion directly contrary to the plain language therein is mistaken. I know beyond the furthest galaxies in this present universe there is very real "outer darkness" because I've been there. So, as far as I can tell, this universe was made in the manner God, Jesus Christ, showed me and therefore has an origin as my notes on mapping space and time indicates.

When I mention God, Jesus Christ, I want people to understand I'm not talking about His humble appearance as a Man some 2000 years ago, on this earth He created and made, I'm saying Jesus Christ exists Eternally in a Pervasive and Transcendent Invisible Way; not JUST His Visible Image that looks like a Man. In other words, God, Jesus Christ, permeates His Creation and exists Infinitely and Eternally beyond His Creation. (no escaping God, He's Every-WHERE and Every-WHEN). Don't turn yourself into an enemy of His.

Again, for particles to define reality that form plasmas, gases, liquids, solids, all elements, reality before our eyes, that respond to intelligent control, irrefutably proves Intelligent Design! The properties of those particles are obviously intelligently designed, there is no other reasonable conclusion based on reality before our eyes! A random anything that explodes, does not all on its own create particles, atoms, chemicals, elements, stars, planets, life, intelligence, etc. etc. so the entire God-denying paradigm (from "big bang" to "the theory of evolution") is ludicrous based on observable reality and the most important facts of history. In other words, it's not like God hasn't shown Himself and proven to us who He is. Jesus Christ stated He is God and He confirms the Sacred Text in the Holy Bible to this very day. In other words, when He came, stated and proved He is the One True God, He told us all how to know Him personally; so, no one has any excuse! NONE!

Look, I could go on and on and on why none of those stories being taught today in the name of science are true, but the truth is I would merely be restating what others have already said and written on the subject. They are being ignored because God-hating people religiously hold onto those cultic beliefs, they would rather believe they came from monkeys and pond scum from a chaotic universe that exists without reason, than that they must be accountable to the One who Commands us all to turn from our wicked ways. In other words, most people hating on Jesus Christ, are the same kind of wicked persons that were destroyed in the past, both in the flood, and in the fire and brimstone that rained down upon Sodom, Gomorrah and the other three cities on the plains of Zoar. Those ashen remains still serve as a warning to us all not to become so wicked that you are marching in the streets flaunting your depravity in front of the rest of us. No, this pride movement of sexual immorality is not a good thing in the least, it spells only doom and destruction for all such persons and perhaps even the ruin of entire cities or whole nations if they embrace such evil, wickedness and overt immorality. Being proud of the very things that brought death and destruction is a sign you are going insane due to demonic possession. You all need to repent immediately and humble yourselves before the Living God, beseeching Him to forgive you, cleanse you from unrighteousness, kick the damned devil out of your lives and keep him out and to give you His Holy Spirit instead. Otherwise, such unrepentant, wicked persons are headed for God's Wrath.

https://www.youtube.com/watch?v=fHQEcqwrYxc

Mathematics - the entire universe, everything we have learned about it to date, relates to mathematics. There are formulas that relate to the properties and motions of the galaxies down to the tiniest particles. Formulas that relate to and calculate the properties and motions of all matter and energy means creation was Intelligently Designed to be intelligently understood. The stars convey Divine Messages and the tiniest particles are used as the building blocks of it all, showing irrefutably Intelligent Design and soundly disproving evolutionary gibberish. (The contents of the Holy Bible are true; stories from people denying the Eternal Creator and His Words therein are not.)

"1 The ancient solar years of the eastern nations consisted of 12 months, and every month of 30 days: and hence came the division of a circle into 360 degrees." - https://publicdomainreview.org/collection/sir-isaac-newton-s-daniel-and-the-apocalypse-1733/

Mathematics is not just applied to all observable phenomenon but is inextricable from the design of creation, proving that when God, Jesus Christ, says He is Supreme in Everything, that doesn't just mean King of kings and Lord of lords, but Mathematician of mathematicians and Scientist of scientists. God is Truth, All True Knowledge comes from Him; anything contrary to His Words (such as the theory of evolution), and the entire evolutionary paradigm of a self-created universe of self-assembling matter and energy, and life forms, is absolute hogwash. Anyone denying the existence of our Eternal Creator, Jesus Christ, has left reality, the plainly recorded facts of history, a sound mind and embraced madness, lies, deceptions and self-delusions.

When you realize that God Himself created and made language and designed letters with Divine Properties and Meanings, that correspond to images, sounds, numbers, musical notes; ALL CREATION and all phenomenon therein; that is when you begin to discover the deeper meanings within the Divinely Inspired Holy Bible.

https://www.youtube.com/watch?v=44FutoWme8A - John 1:1 It is by the Existence of our Eternal Creator and His Words have all things been Created and Made and is why everyone needs to know the contents of the Holy Bible, because His Words are not just the reason for creation, reality, but are interwoven into all things visible and invisible.

https://www.youtube.com/watch?v=g87AKnEMAKo - yes, the Israelites were in Egypt. yes, Joseph was in Egypt. The Holy Bible is verified history; not mythological fiction!

Yes, the letters of the alphabet relate to ancient pictographs. Some in one breath will acknowledge that these letters did, but then in another breath, say that today there is no such association. https://www.youtube.com/watch?v=ph2oxjt0u3w It would be one thing if people were saying that these letters mean something else than the ancient pictographs they came from, but they aren't. They merely point out that these letters, that correspond to ancient pictographs have understandable connotations when applied to the Biblical Text:

https://youtu.be/_3CQtIWnx7c?t=3184 - people are not whimsically assigning nonsense to the scriptures like Dr. Brown seems to imply in his incredulity, but rather are pointing out that these ancient meanings correspond to the text and deepens our understanding of what is written therein. It would be one thing if the ancient beginnings of these letters made no sense whatsoever when applied to the order of these letters in the text, but they do. So, it isn't "nonsense" like Dr. Brown is trying to assert. (I like most of his views and presentations, so I'm not trying to pick on him. Just because we can agree on many things down here, doesn't mean we'll agree on EVERYTHING, at least until the Return of our Lord Jesus Christ in Glory and when He makes us to be Immortal and Glorious like Himself.)

https://youtu.be/g87AKnEMAKo?t=1726 – language, letters and words, pertain to history and reality and since the Holy Bible is Divinely Inspired, it's true throughout time (past, present and future) and beyond! (So yes, people who are applying ancient meanings and connotations to the letters therein are doing so appropriately.)

Astronomy - unbelievers used to cite "Zeitgeist" rather frequently online until believers pointed out that if all ancient civilizations could read the same message in the stars, that was synonymous with the Gospel of Jesus Christ, it only validated the truthfulness of the Holy Bible. God's Message Written in the Stars He Created and Made.

https://www.youtube.com/results?search_query=chuck+missler+signs+in+the+heavens+ - Our Eternal Creator, Jesus, the Christ, tells us plainly in the contents of the Holy Bible that He created and made the stars for specific purposes:

14And God said, "Let there be lights in the expanse of the sky to distinguish between the day and the night, and let them be signs to mark the seasons and days and years. **15**And let them serve as lights in the expanse of the sky to shine upon the earth." And it was so.... https://biblehub.com/genesis/1-14.htm

It was by observing the stars, that the wise men knew the Messiah, God in the flesh, had arrived as He had told His Prophets.

https://www.youtube.com/results?search_query=Yeshua+Mazzaroth - https://www.youtube.com/watch?v=EUQEMqF5dL8 and there are others pointing out the correlation between the celestial and terrestrial - https://bethlehemstar.com/setting-the-stage/

Jesus Christ, the One True God, Created and Made the Heavens and the earth and so He tells us who He is in both.

Psalms 19:1-4 King James Version (KJV)

The heavens declare the glory of God; And the firmament sheweth his handywork. Day unto day uttereth speech, And night unto night sheweth knowledge. There is no speech nor language, Where their voice is not heard. Their line is gone out through all the earth, And their words to the end of the world.

Which is how the wise looked up and saw His Sign/Star announcing His Arrival on earth and why they journeyed to find and worship the Living God. (Watch the cited presentations to see how the stars announced God in the flesh at the moment of His Arrival as He had told His Prophets in the contents of the Holy Bible.)

Yes, the many hundreds of prophecies about God coming in the flesh as the Messiah were compiled centuries before He did so; so yes, Jesus Christ is the One True God; no ifs, ands or buts!

The Law and Prophecies given us from God, Jesus Christ about Himself were in fact compiled centuries prior to His Advent in the flesh on earth, this world He Created and Made.

https://biblearchaeology.org/research/new-testament-era/4022-a-brief-history-of-the-septuagint - history records and is confirmed that the contents of the Old Testament existed centuries before God Almighty, known in the English language as Jesus Christ, came in the flesh!

https://www.accordingtothescriptures.org/prophecy/353prophecies.html hundreds of prophecies God, Jesus Christ, told His Prophets about Himself; fulfilled when He arrived, the rest of the prophecies Jesus Christ, the One True God, gave us in the contents of the Holy Bible, He is still fulfilling to this very day.

God, Jesus Christ, has been fulfilling His Words precisely as He has told us in the contents of the Holy

Bible, some of those prophecies were fulfilled even down to the exact moment in time (day) He foretold:

https://youtu.be/o8AXk-zW-nQ?t=3092

Some falsely accuse the Holy Bible of being self-fulfilling, rather than being provably and verifiably Divinely Inspired (clearly given us from our Eternal Creator who has been accurately telling us about the future in the contents of the Holy Bible.)

So, then I tell them about the Divine prophecies in the Holy Bible that have happened in recent history before the eyes of the world, prophecies such as smart missiles, jets, modern automobiles, the reformation of the nation of Israel in the land of ancient Israel, wars and conflicts, and much, much more:

http://www.blastthetrumpet.org/PublicLetters/AAAUpdatedPublicAlertsMattersofLifeandDeath/Prophecies%20Show%20Divine%20Inspiration%20of%20the%20Holy%20Bibl2.pdf - yes, we still have ancient manuscripts proving those prophecies about even modern events and technologies were written over two thousand years ago. - https://carm.org/about-the-bible/manuscript-evidence-for-superior-new-testament-reliability/ and again, artifacts and other evidence for the facts of history that some of those prophecies were given closer to three thousand years ago.

https://biblearchaeologyreport.com/2019/01/12/top-ten-discoveries-in-biblical-archaeology-relating-to-the-old-testament/ - artifacts and ancient sites buried in the earth continue to be discovered to this very day, validating the historicity of the ancient text of the Holy Bible. https://www.google.com/search?client=opera&q=stele+oldest+artifact+confirming+Bible&sourceid=opera&ie=UTF-8&oe=UTF-8#ip=1 - there is an abundance of very real ancient sites and artifacts that confirm the historicity of the Holy Bible.

So much external evidence supports the veracity of the Divinely Inspired Sacred Text in the Holy Bible, that people spend entire lifetimes just examining a few of those aspects, let alone all of them. (The Holy Bible has generated billions of books, research papers on and offline throughout history to this day, such that no one but God Himself knows all the facts about His Words therein.)

http://www.blastthetrumpet.org/PublicLetters/AAAUpdatedPublicAlertsMattersofLifeandDeath/Updates053016/NO

%20MORE%20BRAINWASHING%20p1.pdf - the historicity and veracity of the contents of the Holy Bible have been thoroughly established. Anyone referring to those contents as fiction, myth or legends, is lacking an enormous amount of truthful information.

http://www.blastthetrumpet.org/PublicLetters/AAAUpdatedPublicAlertsMattersofLifeandDeath/Updates053016/NO%20MORE%20BRAINWASHING%20p2.pdf - the more mankind learns and discovers about history and reality, the more that knowledge only validates the provably and verifiably Divinely Inspired contents of the Holy Bible.

In other words, all true science/knowledge confirms that God Almighty, Jesus Christ, is telling us all the Truth in the Holy Bible about past, present and future, about things visible and invisible, about things physical and spiritual, things temporal and things eternal, and even about very real places departed spirits (people who die) go to after death. So, make certain you have obeyed the Commandment of Jesus Christ, the One True God, to Repent of every evil thought, word, way and deed and Get Baptized in His Name and pray to receive His Own Holy Spirit of Truth, whereby you will KNOW and LEARN from the One True God; now and forever. Then study His Instructions to Live By, in the Holy Bible, and apply them accordingly.

https://www.facebook.com/photo.php?fbid=6609426679136235&set=pb.100002069048072.-2207520000&type=3

You don't have to be a rocket scientist or professional geologist to know solid rocks don't fold without breaking; so, these sedimentary layers with marine fossils in them are evidence of the flood; not billions or millions of years:

https://assets.answersingenesis.org/doc/articles/pdf-versions/rock-layers.pdf

https://www.google.com/search?client=opera&hs=9iC&sca_esv=934b7f5af1c0dd08&sxsrf=ACQVn0_E091c1iF1wt1G1SPERKWEsFVFPg:1709399776050&q=folded+rock+layers&tbm=isch&source=lnms&sa=X&ved=2ahUKEwivxrizitaEAxWBOTQIHX5CANMQ0pQJegQICxAB&biw=1240&bih=618&dpr=1.5

https://creationontrial.com/articles/folded-rocks/ - people should be questioning deep time claims regarding the sedimentary layers in which fossils of all manner of creatures, from single-celled organisms to dinosaurs, have been found, because there are many reasons those claims are NOT TRUE! NEVER believe STORIES about pre-history OVER the FACTS of history! Especially the well verified history in the provably Divinely Inspired Holy Bible!

The most common way to form sedimentary layers and fossils in those layers is by water, FLOOD! yet due to the fact so many are trying to hate on God and the Truth in the Holy Bible THEY ARE MAKING UP STORIES! LIES! Just because you can fool people with lies, is no reason to dismiss the TRUTH!

I visited https://www.nps.gov/dino/planyourvisit/index.htm notice that the story of "millions" of imagined years doesn't jive at all with the plainly observed facts in the display: https://www.nps.gov/dino/planyourvisit/quarry-exhibit-hall.htm Claiming that a river periodically buried the bones of so many enormous creatures over "millions" of imagined years is ludicrous. Look! at those sediments the bones are all buried in! There's no evidence whatsoever of seasonal flooding, layer by layer over millions of years! INSTEAD, we see a HUGE variety of bones and fossils of all kinds ALL JUMBLED TOGETHER in meters thick common sediments! FLOOD! and not seasonal river flooding over "millions" of imagined years, but MASSIVE FLOOD; enough to move 20-40 feet worth of sediments or more at once and bury all those massive creatures!

https://www.google.com/search?client=opera&hs=4cs&sca_esv=df525f6b0b574dfa&sxsrf=ACQVn09Yt0CruSgzHxZ0Z1kG9zG1Pr2bfg:1709401965474&q=fossil+graveyards&tbm=isch&source=lnms&sa=X&ved=2ahUKEwictLjHktaEAxV6HDQIHdVEC9IQ0pQJegQICxAB&biw=1240&bih=618&dpr=1.5 - piles of fossilized bones all buried together worldwide are best explained by a flood! God-deniers, Biblical flood deniers, GO OUT OF THEIR WAY to deny that fact! Stop listening to LIES just because someone telling those lies claims to be some kind of "expert"; instead, BELIEVE YOUR OWN EYES! THINK FOR YOURSELF!

Geology and the fossils of fully formed creatures the vast majority of which are all marine organisms with no examples of any morphing from bacteria to dinosaurs are no friend to evolutionary claims whatsoever! THINK! why are bacteria still bacteria today if "billions" of imagined years ago, they "evolved" into more complex organisms? THINK! how many times IMPOSSIBILITIES must be embraced to believe their STORIES! like over and over and over again, ENOUGH complex beneficial mutations HAD TO SURVIVE in order to go from asexual reproduction to sexual reproduction AND THOSE COMPLEX MUTATIONS had to occur SIMULTANEOUSLY in order to have both males and females. MILLIONS of times OVER, there had to have been COMPLEX ADDITIONS, BENEFICIAL MUTATIONS, EVOLUTIONS, (massive amounts of additional genetic coding, DNA, specific ordered molecules, that produced added structure to the formerly single celled organism) occurring in order to have their story of macroevolution "goo to you by way of the zoo", ABSOLUTE NONSENSE. Despite digging the world over they have never found evidence for such madness, the "Cambrian explosion" is dubbed such because of that lack of evidence. "macro-evolution" of one kind of creature into a completely different kind of creature DIDN'T HAPPEN! one cell in the past did not become bananas and the monkeys that eat them no matter how many imaginary years go by, IT'S IMPOSSIBLE! no ifs, ands, or buts! When we see fully formed complex organisms buried in sedimentary layers worldwide, single celled organisms to multi-trillion celled dinosaurs, with over 95% of those fossils being marine, it's overwhelming evidence that macroevolution is HOGWASH! Instead, it's all evidence of the historicity of the Biblical Creation account.

You have to THINK about the EVIDENCE, instead of believing anyone else's faulty opinions about that evidence. When you put deep time CLAIMS against the well verified history in the Holy Bible, the Holy Bible is telling us the TRUTH; NOT THEIR IMAGINARY CLAIMS!

https://www.facebook.com/photo/?fbid=6382847548460817&set=a.115635768515391

1) SOMETHING FROM NOTHING IS IMPOSSIBLE! - so everyone denying our Eternal Creator is demonstrating they are willfully ignorant and self-deluded! The universe did not come from NOTHING! The universe did not create itself! Logical impossibilities like SOMETHING FROM NOTHING is FALSE! for ANYTHING FINITE to exist (like this present universe) it is absolutely evidence of our ETERNAL CREATOR!

2) NO ONE GETS A PASS TO ENTER VAIN IMAGINATIONS AS "SCIENCE" IF THEY ARE SKIPPING OVER FACT NUMBER 1! IF PEOPLE DO NOT ACKNOWLEDGE OUR ETERNAL CREATOR, THEY ARE NOT A SCIENTIST! NOT SOMEONE WHO IS RATIONAL, REASONABLE OR EVIDENCE BASED AT ALL!!!!!!!! So, NO! get it through your presently deceived minds! NO! the universe from nothing is NOT science! NEVER WILL BE SCIENCE! it's just willfully ignorant, deceived and self-deluded persons putting forth their impossible imaginations in the name of science and the scientific community SHOULD BE TELLING THOSE PEOPLE TO GO BACK TO GRADE SCHOOL AND DEMAND A PROPER EDUCATION! One that is based on the well verified TRUTH in the Holy Bible!

So, the second impossibility evolutions are embracing, is that the universe from nothing, assembled itself during the "big bang" cosmic inflation event. In other words, SINCE THEY HAVE ENTERED THE REALM OF SELF-DELUSION, THEY HAVE TO COME UP WITH SOME KIND OF STORY ABOUT THEIR IMAGINARY DEEP TIME PRE-HISTORY and so embrace another IMPOSSIBILITY that order came out of chaos, knowledge came out of ignorance and all forms of matter and energy assembled itself to create, particles, atoms, elements, chemicals, stars, planets, space and time, REALITY FROM NOTHING CREATED ITSELF! Everyone should be telling these people posing as "scientists" not just NO, but NEVER! not in any amount of imaginary time did the universe from nothing create and assemble itself! NEVER! STEP DOWN! RIGHT NOW! you and your delusions have injured enough people already! you are NOT a scientist if you embrace madness!

People you have to trust your own perceptions when it comes to REALITY and why REALITY exists! NOTHING IN ALL CREATION ASSEMBLES ITSELF, NOT ONE PARTICLE IN THE ENTIRE UNIVERSE CAME FROM NOTHING, DEVELOPED COMPLEX PROPERTIES, FORMED ALL THE ASPECTS OF WHAT WE OBSERVE AS REALITY! EVERYTHING CREATED AND MADE REQUIRES MASSIVE AMOUNTS OF ENGINEERING, DELIBERATE INTENTIONAL DESIGN, AND DESPITE INTELLIGENCE, DELIBERATE APPLIED ENERGY TO MATTER TO BUILD ANYTHING, IT STILL

DECAYS! STILL IS SUFFERING FROM ENTROPY! So, NO! you do NOT get a pass on OBSERVED SCIENTIFIC LAWS to claim that those laws did not exist during the "big bang" cosmic inflation, but only came after the universe from nothing assembled itself in all its complexity.

CHEMICAL, STELLAR AND COSMIC EVOLUTION AS CURRENTLY TAUGHT TODAY IS NOT SCIENCE! AGAIN, IT WILL NEVER BE SCIENCE! NOT IN ANY IMAGINARY AMOUNT OF TIME! IMPOSSIBLE! pitch that crap as the crap that it is!

I've written on this topic enough so that anyone with basic reading comprehension can understand that the entire evolutionary paradigm currently wrongly being taught as science is absolutely NOT science, it doesn't even rise to the merit of science fiction, BECAUSE IT IS IMPOSSIBLE by everything we know with all certainty to be true.

(Like the universe came from our Eternal Creator, who has plainly said so, and showed Himself and proved Himself, by speaking and changing reality with His Words! JESUS CHRIST IS GOD! no ifs, ands or buts!) The moment anyone denies Him, is the moment they immediately show themselves to be willfully ignorant, deceived and self-deluded!

NONE of the sciences support the God-denying story being wrongly taught in the name of science currently; NONE! **If you want to be taken seriously as a scientist, YOU MUST MENTION OUR ETERNAL CREATOR AND HIS WORDS IN THE HOLY BIBLE! If you have a theory about the origin of the universe and the origin of life, IT MUST STAND UPON WHAT GOD HIMSELF TELLS US IN THE HOLY BIBLE!**

Take astrophysics, cosmology, for example. People are pointing their cameras at a super-massive galactic core, burning and shining brightly and saying, "Look, a 'black hole'!" I'm looking and thinking, these people who believe everything came from nothing, a universe that assembled itself in all its complexity, are at least consistent in their madness; such that an enormous amount of mass and energy is in their mind just the opposite - a "black hole".

So modern genetics finally determines everyone living today come from one mother, whom they've dubbed "Mitochondrial Eve", and rather than say, WOW! the Holy Bible has told us for millenniums that Eve is the "mother of all living", they actually still laugh at those contents. And then when astrophysicists discover the universe is expanding, the galaxies and stars moving apart, they would rather imagine, "dark matter" and "dark energy" is the cause, instead of just again saying, WOW! God tells us in the Holy Bible (throughout recorded history) that He stretches out the heavens!

LOOK! EVERYTHING WE HAVE OBSERVED ABOUT CREATION VERIFIES THAT THE ONE TRUE GOD, OUR ETERNAL CREATOR, HAS BEEN TELLING US THE TRUTH LONG BEFORE SCIENCE HAS EVEN BEEN ABLE TO VERIFY IT! And yet, when I say, the Holy Bible is not just provably true, not just provably and verifiably Divinely Inspired; but yes! is indeed a Book of Science, I still get people laughing; acting like they have not one clue either about reality or what God plainly tells us in those contents!

If I had decades to go on writing, I could do so! ALL OF THE SCIENCES, ALL TRUE KNOWLEDGE PLAINLY AND CLEARLY VERIFIES THE CONTENTS OF THE HOLY BIBLE, BECAUSE THE ONE TRUE GOD, JESUS CHRIST, DOESN'T LIE! THOSE CONTENTS ARE ABSOLUTELY TRUE!

Conversely, NONE of the sciences support the God-denying madness being taught currently as the evolutionary paradigm of a universe from nothing that assembled itself and the magical morphing cell that sprang into existence and created all living creatures; NONE! It's all total rubbish, to put it bluntly. And yes, their story really is that ridiculously impossible, even though when I break it down into easily understandable terms, they habitually repeat their brainwashing back to me... no, you just don't understand... YES, I DO! I understand you parrot your terms as if you want to baffle me with them, when the truth is, "evolution" is indeed a story about "goo to you, by way of the zoo". And LAUGHABLE, despite: "alleles, epigenetics, DNA, RNA, genomes, etc." or any other words you toss into the salad of evolutionary insanity to try and make it seem reasonable or defensible, when it isn't and never will be, **BECAUSE I HAVE SHOWN YOU PLAINLY THAT ENTIRE STORY IS NOT SCIENCE! IT'S NOT EVEN SCIENCE FICTION! IT'S IMPOSSIBLE!!!!!!!!**

An imagined "singularity" that came from nothing didn't explode to create space, time, all forms of matter and energy; our universe. A cosmic soup of particles didn't assemble themselves into the elements and chemicals, then gradually come together to form stars and planets, then start spinning all on their own, then create a primordial soup on earth, then create life, ... NONE of that happened! LUDICROUS NONSENSE! pointing to a super-massive burning object is not a "black hole". And denying God, doesn't give you the right to imagine "dark matter" and "dark energy", invisible things as your explanation while you laugh at people who acknowledge God Almighty, merely because you haven't seen Him. In other words, your story of creation and life actually takes FAR MORE "blind belief" than the well verified Biblical Creation account that tells us plainly reality exists because our Eternal Creator exists. And further tells us that the universe was created and made by Him, merely by thinking of it, and speaking it into existence.

"Let there be Light!" the first chapter of Genesis. Now some want to limit the Holy Bible and say surely God wasn't referring to ALL aspects of Light, spiritually and physically, but yes! REALITY IS DEFINED BY HIM, so Light in all its aspects is the manifestation of His Creative Imagination, Divine Thoughts! Ultimately His Divine Thoughts, trump all our vain imaginations, so ultimately His Creation will be exactly as He intends, harmoniously existing with Him forever. IN THE MEANTIME, people He created and made are imagining and doing evil things and that is the reason for evil in this world, and this present fallen creation being subjected to decay/entropy. (evil is allowed by God for a limited amount of time, but will one day end, just like this fallen heavens and earth.)

This current earth has been cursed (is under Divine Consequences on account of the devil, and our own evil imaginations, words and deeds). These consequences are mentioned in the Holy Bible with one of the lists of them being found in Deuteronomy 28. Essentially, painful labor, weeds, thorns, plagues, pestilences, diseases, suffering, decay, destruction and death has come upon this present heavens and earth, due to EVIL that was conceived first in the thoughts of the devil (Lucifer fell and became known as satanil, adversary of God, Jesus, the Christ). But also, all those consequences came upon mankind as a whole because we all have sinned, (imagined evil and did evil at some moment in our existence).

For the Saved, the Redeemed, those Grateful for the Love, Mercy and Forgiveness of our Eternal Creator, and willing to perfected by Him, He tells us that He indeed will make us to be Perfect and Holy, Inherently Righteous and Good, like Himself. But for the wicked, who refuse to repent of their wicked ways, He tells us they will all be cast into the lake of fire and prevented from ever again harming His Children (those who consent to be perfected by Him and want to be Good and Righteous like Him forever). Rev. 21:7,8 and 1 Cor 15

So, Creation is indeed evidence of our Eternal Creator, and He is indeed the God who plainly tells us all who He is and how to know Him in the contents of the Holy Bible. **We need to put the Holy Bible back in our schools immediately and remove all those lying text books** of nonsensical gibberish, before we have even more ignorant, deceived and self-deluded fools who actually imagine that evil pile of crap called evolution is science.

https://www.youtube.com/watch?v=GjvuwneORrE - Trey Smith does proper respect to the theory of evil-u-shun and the evidence of Genesis - https://www.youtube.com/watch?v=GjvuwneORrE

https://www.youtube.com/watch?v=UM82qxxskZE - this presentation also does "Genesis Is It History". Watch their entire series on the subject, not just this movie.

Yes! Genesis is redundantly proven history and no one dismissing it or the contents of the Holy Bible belongs in public education! WE MUST RETURN THE HOLY BIBLE BACK WHERE IT BELONGS; THE HOLY BIBLE IN PUBLIC EDUCATION! And no more evolutionary nonsense!

IF YOU HAVEN'T GRASPED THAT THE ENTIRE EVOLUTIONARY PARADIGM INCLUDING THE THEORY OF

EVOLUTION HAS BEEN SOUNDLY DISPROVEN, THEN READ THIS AND ALL CITATIONS IN THIS BOOK AGAIN! LET IT SINK IN! **WHAT IS BEING TAUGHT TODAY IN THE NAME OF SCIENCE IS NOT SCIENCE! THEIR STORY FROM BIG BANG TO GOO TO YOU BY WAY OF THE ZOO IS NOTHING MORE THAN A STEAMING PILE OF IMAGINARY CRAP! CRAP THAT HAS CAUSED MANKIND TO WRONGLY DISMISS THE TRUTH CONTAINED IN THE HOLY BIBLE! SO, THEIR STORY IS EVIL! LIES FROM THE DEVIL! STOP BELIEVING THAT CRAP!**

Look, the moment anyone builds their ideas upon the universe from nothing, is the moment they are no longer in the realm of science at all and instead are lost in a world of lies, deceptions and self-delusions! **YOU MUST BEGIN WITH OUR ETERNAL CREATOR AS THE CAUSE FOR REALITY; THAT IS THE ONLY RATIONAL EXPLANATION! AND HE PROVED HE IS THE GOD WHO DECLARES HIMSELF TO US IN THE HOLY BIBLE!** I am pulling the rug of the universe from nothing out from under the rest of their entire errant ideas taught wrongly as science currently! **NO! the moment anyone denies our Eternal Creator, is the moment they are no longer a rational, reasonable person; most definitely NOT a "scientist"!**

For science to be science, it must be built upon the foundation of truth, otherwise, it isn't science! The universe exists because of our Eternal Creator. THAT IS THE TRUTH! That Eternal Creator is JESUS CHRIST as a FACT OF HISTORY, AND OBSERVABLE REALITY! So let me be crystal clear, that the moment anyone strays from the Divine Revelation of Truth given us by God Almighty Himself, is the moment they have erred, whether they are a scientist or not! YOU MUST STAY WITHIN THE FRAMEWORK OF PROVABLE AND VERIFIABLE TRUTH FOR YOUR "THEORIES" TO BE IN ANY WAY SCIENTIFIC! You must stay within the crystal clear provably and verifiably Divinely Inspired Words of God Almighty Himself given us in the Holy Bible!

Anyone disdaining the Holy Bible, does not know God! does not know Truth! and should be ignored! **The Holy Bible is provably and verifiably Divinely Inspired, ROCK SOLID TRUTH!!!** If you don't know that you need to study it and pray to our Good Lord Jesus Christ to give you His Understanding of all things. If you're thinking knowledge in your head is why you dismiss the contents of the Holy Bible, THEN THAT KNOWLEDGE IN YOUR HEAD ISN'T TRUE... you are believing lies.

(Read The Holy Bible is Provably and Verifiably Divinely Inspired AGAIN! Until you KNOW ABSOLUTE TRUTH WITH ALL CERTAINTY BECAUSE YOU KNOW JESUS CHRIST, OUR ETERNAL CREATOR; THE ONE WHO TOLD US HE AND HIS WORDS IN THE HOLY BIBLE ARE TRUTH!)

Far too many people today have sadly been so thoroughly deceived by coerced indoctrination, public schools, that I am very concerned about the fate of our nation and the world due to such massive ignorance, deception and self-delusions. There are people who are actually seriously unable to tell the difference between the contents of the Holy Bible and known works of pure fantasy like a comic book, or ones that wrongly imagine we are arguing that just because the Holy Bible mentions God, Jesus Christ, that we are saying it makes God to be real. (they're not grasping any of the Divine Proofs of the scriptures, it's as if they don't understand logic, reason, and can draw reasonable conclusions from verifiable, accurate information.) Their lame perspectives are so incredibly limited that they then say things like, "I suppose spider-man is real also". IN OTHER WORDS, EDUCATION HAS COMPLETELY FAILED ALL THESE PEOPLE IN THAT THEY ARE TOTALLY UNABLE TO DISTINGUISH FACTS FROM FICTION, HISTORY FROM FANTASY, TRUTH FROM LIES, OR OUR VERY REAL ETERNAL CREATOR from some overtly purely imaginary anyone or anything.

I, most alarmingly and disturbingly, find myself encountering vast numbers of such people who actually imagine they are presenting themselves as intelligent by mocking the contents of the Holy Bible (that virtually all of them have never even read, let alone seriously studied.

Instead, sometimes I'm actually thinking that they've been so abused with so many lies taught them in our public schools that they're acting not just ignorant, deceived, brainwashed and self-deluded, but as if they've lost the ability to discern reality from hallucinations, imaginations, delusions; to the point they're acting as if they've suffered brain damage, a lack of ability to comprehend simple sentences, plain facts, etc.

As a result, I find myself routinely posting replies like the following:

I've lately encountered people who equate the Holy Bible with a comic book or any other religious text and when I encounter such people, I think these things:

1) they must not have read the Holy Bible or if they have, they have failed to understand it completely

2) if they cannot tell the difference between historic literature and overt fantasy/fiction they must be so incredibly mentally challenged that it's unlikely they will ever come out of their state of ignorance and delusion, at least until they stand before God Almighty and are judged

by Him

3) but I must attempt to reason with the unreasonable regardless, because the Almighty Commands those who know Him to do so

So more than likely, I am not going to persuade you as to why your post (one of the many today of people attempting in vain to denigrate the Holy Bible) demonstrates a lack of the most important information on earth, but it does.

The Holy Bible is unlike any other book in all the world in that it is provably and verifiably Divinely Inspired.

And yes, I have studied enough of the major worldviews on earth (atheism, evolutionary cultic believers, muslims, hindus, buddhists, along with those who affirm the Holy Bible) and I can tell you having examined them all with their respective beliefs that only the Holy Bible is supported by accurate knowledge and is rational and reasonable, all other worldviews are indefensible.

I can say that because the Holy Bible not only tells us the facts about reality past, present and future, but because it tells us all who the One True God is AND HOW TO KNOW HIM PERSONALLY BEYOND ALL DOUBT. So, the Holy Bible is VERIFIABLY DIVINELY INSPIRED because the GOD of the Holy Bible, JESUS CHRIST, tells us all how to KNOW HIM right here and now (and forever) and how to learn from Him personally; so, no one has any excuse.

Jesus Christ is the One True God and His Words in the Holy Bible forever tower above all other literature. His Words that He has made public in the contents of the Holy Bible forever test and try all other opinions from ANYONE past, present or future! Anyone ignoring the Words of God in the Holy Bible, should be ignored by all the rest of humanity! (they are demonstrating INEXCUSABLE, willful ignorance and rebellion against the Almighty!)

https://www.facebook.com/photo.php?fbid=4890242114388042&set=pb.100002069048072.-2207520000&type=3 - the Holy Bible unlike any other book on earth has been accurate historically, scientifically, not just about the past and present but about the future throughout the history of mankind! many hundreds of detailed prophecies that include things like modern cars, jets, missiles, specific wars, etc. are found in words that were written over 2500 years ago! So, NONE of the other religious texts on earth can compare to the Holy Bible, not even close!

My friends, if you had all been properly educated, you would have been told you can KNOW and LEARN FROM our ETERNAL CREATOR, JESUS, the Christ, PERSONALLY! He is the Truth the ONE WHO DOES NOT LIE, and is how you will not be deceived by lies like evolution or anything else! Now that you have been told, this is what you must do to know and learn from Him:

1) Obey His Commandment given to all mankind: REPENT! turn away from every evil thought, word, way and deed. REPENT! stop imagining you know better than the One who Created and Made you! Learn to listen to and obey the Living God, our Heavenly Father, our Lord and Savior, JESUS, the Christ!

2) Go and get baptized in His Name, by someone who can look you in the eyes and tell you they KNOW the One True God, JESUS, the Christ! And can baptize you in His Name properly by total immersion. Right hand upon your forehead, baptized in the Name of our Lord Jesus Christ. You should be praying to receive His Holy Spirit, the Spirit of Truth, the Spirit of Christ, the Spirit of the Living God!

When God, Jesus Christ comes upon and within you by His Holy Spirit, you will KNOW Him!

3) Study His Words in the Holy Bible and apply them to your life always and forever! Never stop learning from God, never stop seeking to know Him more and more! In the Holy and Omnipotent Name of our Lord Jesus Christ, May He Open your eyes to see, and your ears to hear Him speaking to you!

Once you KNOW the Living God, tell everyone who still doesn't know the Almighty, that they NEED to KNOW Him, the quality of their life and their everlasting destiny depends on it!

Made in United States
Cleveland, OH
06 November 2026

47664314R00254